Windmills in Lambeth

an historical survey

by Michael Short

London Borough of Lambeth
1971

Published by the Director of Library and Amenity Services of the London Borough of Lambeth, Tate Central Library, Brixton Oval, London SW2

Designed by Adrian Hodgkins
Printed by Lund Humphries

ISBN 0 9501893 0 8

Contents

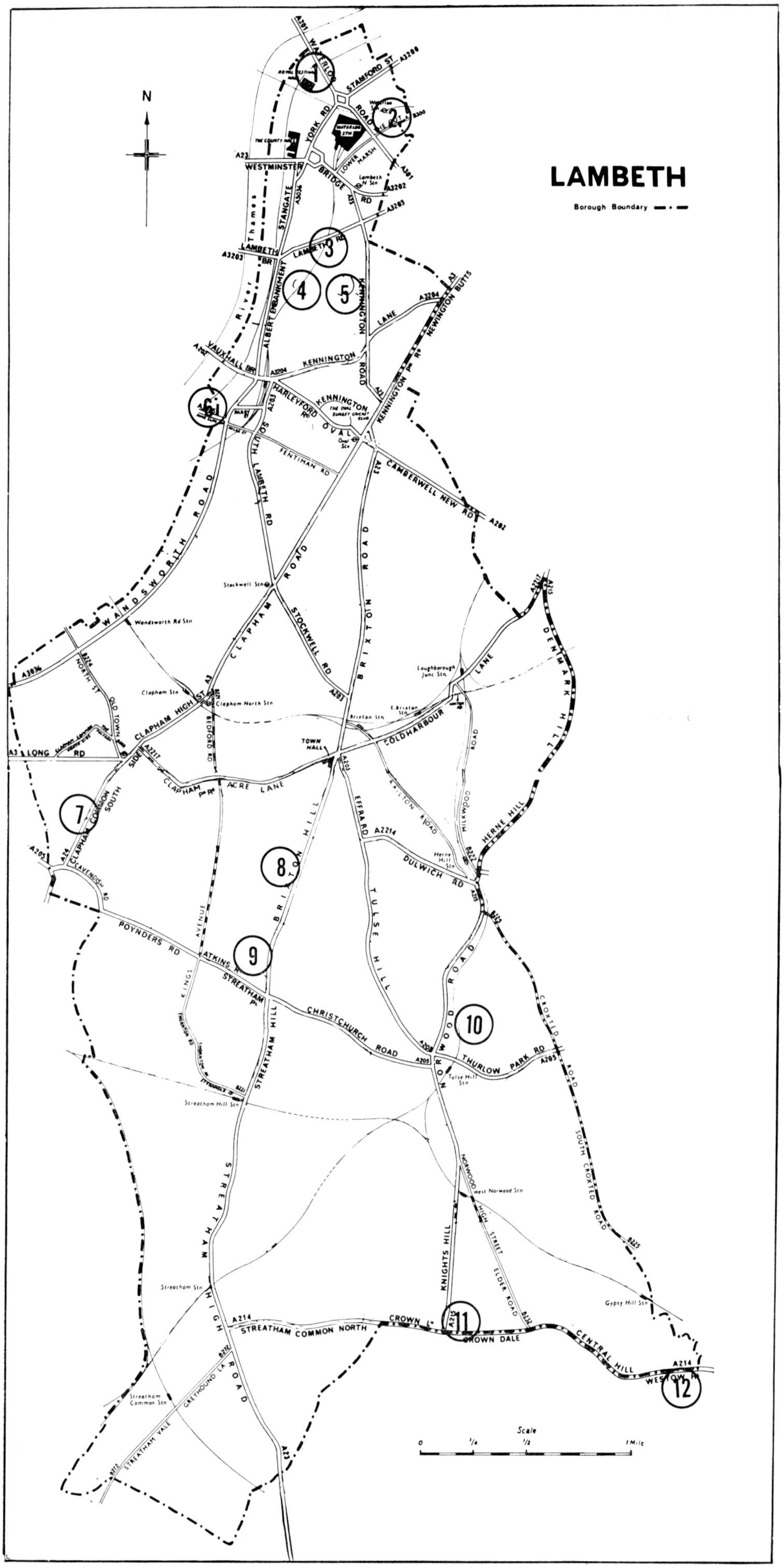

1. Map of the Lambeth area, showing the sites of windmills mentioned in this survey.

List of illustrations

17. 'Windmill in Mill Street, Lambeth': a sepia drawing by John Buckler, dated 1828. *Duchy of Cornwall Office.*

18. Detail from a view of London and Westminster by T. Nugent, dated 1820, showing in a rather crude fashion the mills at sites 3 and 4. *British Museum.*

19. A watercolour entitled 'Lambeth Mill'. *Lambeth Public Libraries.*

20. 'Orchard's Flour Mill, Lambeth': a sepia drawing by John Buckler, dated 1828. *Duchy of Cornwall Office.*

21. A watercolour by Paul Sandby of about 1780, showing the Lambeth 'Drug' mill. *Guildhall Library.*

22. An engraving by Paul Sandby, based on the previous view. *Greater London Council.*

23. A watercolour entitled 'The Mill belonging to the Apothecaries Company of Lambeth'. *Guildhall Library.*

24. An ink and watercolour painting showing the 'Drug' mill from the opposite side. The mill at site 3 can be seen in the background. *Lambeth Public Libraries.*

25. A nineteenth-century engraving, clearly derived from Sandby's original watercolour. *From Walford's* Old and New London.[B30]

26. 'View of the Thames near Vauxhall': a watercolour by John Cleveley, showing Randall's mill at Nine Elms. *Private collection.*

27. Detail from an eighteenth-century 'View of the River from Vauxhall Stairs' by W. Marlow, showing Randall's mill and associated buildings. *Lambeth Public Libraries.*

28. Randall's mill as depicted in the 1808 Perambulation of the Parish of Lambeth. *Lambeth Public Libraries.*

29. An engraving of Randall's mill, from a drawing by S. Owen, *first published in May 1809. Lambeth Public Libraries.*

30. 'Near Vauxhall': an etching by Samuel Prout, dated October 1815. *British Museum.*

31. 'Near Vauxhall Bridge': a lithograph by F. E. Martens, published in October 1829. *Lambeth Public Libraries.*

32. Detail from Leigh's 'Panorama of the Thames' of *c.*1830, showing the high fantail fitted to Randall's mill. *Lambeth Public Libraries.*

33. 'Mill near Vauxhall – study from nature': a watercolour by J. Varley, dated July 1830. *British Museum.*

34. Another view of Randall's mill by J. Varley, showing the spring shutter sails. *Brighton Corporation.*

35. A watercolour by Thomas Hollis, inscribed 'Apothecaries mill, Nine Elms'. *Guildhall Library.*

36. 'Plan for the Tubbs at Vauxhall China mill': a drawing amongst the papers of John Smeaton. *Royal Society.*

37. 'Plan for the Tub machinery at the China Mill – the Design Executed': a subsequent drawing by Smeaton. *Royal Society.*

38. Design for a set of common sails by John Smeaton. *Royal Society.*

39. 'A Projection for Weathering Windmill Sails in the manner of those at Vauxhall' by John Smeaton. *Royal Society.*

40. Schedule attached to a lease of 1651 relating to Clapham windmill, listing the 'goods things implements and utensills in and about the said Mill'. *Lambeth Public Libraries.*

41. Ashby's mill, Brixton: an early watercolour. *Lambeth Public Libraries.*

42. Ashby's mill before restoration, showing the gallery still in position round the cap. *Lambeth Public Libraries.*

43. A nineteenth-century sketch of Ashby's mill, in which the artist has 'replaced' the sails for the sake of greater realism. *Lambeth Public Libraries.*

44. A general view of the mill yard, now laid out as a public garden. *Greater London Council.*

45. A cross-section of the mill before restoration. *Greater London Council.*

46. A cross-section after restoration, showing the sails and internal machinery. *Greater London Council.*

47. A leaflet published about 1914, showing the prices of various products supplied by Joshua Ashby & Sons Ltd. *Greater London Council.*

48. A photograph of Ashby's mill taken before the original sails were removed in 1864. *From the collection of the late M. T. Mason.*

49. Ashby's mill as it appears today.

50. Rear view of the cap.

51. One of the six centring wheels which keep the cap in position.

52. The rack mechanism which enabled the cap to be turned into the wind.

53. The method of attaching the sails to the windshaft by means of a canister, or poll-end.

54. An early pencil sketch of Ashby's mill, showing the stage at first-floor level and the adjacent brick-fields. *Lambeth Public Libraries.*

55. The brake wheel, which engages with the main vertical shaft.

Acknowledgments

This survey is an extension of the research carried out by K. G. Farries and the late M. T. Mason and published in their monumental work *The Windmills of Surrey and Inner London*.[B19] Owing to the large area covered by that volume, it was not possible to deal with each locality in depth, especially an area such as Lambeth, where the available information is both confused and contradictory, and therefore I have attempted to fill in gaps in the history of windmills in Lambeth and to extend the detail of the survey by obtaining as much further information as possible. Without the work of Farries and Mason as a basis and as an example of painstaking scholarship, the present survey could never have been undertaken. Moreover, Mr Farries has been most generous in making available for consultation all the notes, drawings and photographs in his own collections, despite the possibility of different conclusions being arrived at from the available evidence, and thanks are due to him for his encouragement, suggestions and advice.

Especial thanks are also due to the Archivist of the London Borough of Lambeth, Miss M. Y. Williams, and her Assistant Archivist, Miss R. Edwards, who have allowed me free access to the Surrey Collection and Archives at the Minet branch of Lambeth Public Libraries, despite the large amount of inconvenience caused to them by my use of their office as a private reading room and my persistent questioning on obscure points of local history.

In addition, thanks are due to the staff of the institutions listed below for making available the materials in their custody in an efficient and helpful manner:

Battersea District Library (London Borough of Wandsworth)

British Museum Reading Room, Map Room, Print Room, State Paper Room, and Newspaper Library, Colindale

British Transport Historical Records

Croydon Central Reference Library

Department of the Environment Library, Lambeth Bridge House

Duchy of Cornwall Office

Greater London Council Record Office, Photograph Library, Map and Print Collections, Historic Buildings Section, Legal Department, and Lambeth District Surveyor

Guildhall Library

Kent County Council Archives Office, Maidstone

Lambeth Central Reference Library

Lambeth Palace Library

London Museum Library

National Maritime Museum Library

Public Record Office

Royal Society Library

Society for the Protection of Ancient Buildings

Surrey Record Office, Kingston-upon-Thames

Victoria and Albert Museum Print Room

I would also like to thank the following individuals who have supplied information or otherwise contributed towards this survey:

Mrs Parsons, Westow Hill

J. E. Ruch, Esq., Montreal, Canada

A. W. Scott, Esq., West Norwood

The Earl Spencer, Althorp, Northants

Alan R. Warwick, Esq., Upper Norwood

Finally, acknowledgement is gratefully made to the persons and institutions listed below for permission to make use of photographs of documents, paintings, prints and drawings in their possession:

Brighton Corporation

Trustees of the British Museum

Duchy of Cornwall

K. G. Farries, Esq.

Greater London Council

Guildhall Library

London Borough of Lambeth

Royal Society

Victoria and Albert Museum

Introduction

The aim of this book is to present a survey of all the windmills which have existed in the Lambeth area, giving as far as possible complete descriptions of them and their sites, and information about the people who worked in them. The area covered is that of the present (1970) London Borough of Lambeth, which serves as a convenient demarcation although it is quite an arbitrary area as far as windmills are concerned. Two of the sites surveyed are in fact just outside the borough boundary, but have been included as being of interest to Lambeth residents.

As far as possible, information has been obtained from original sources rather than from printed publications, as the accuracy of local history books, and also, regrettably, windmill books, leaves a great deal to be desired as regards the Lambeth area. Accordingly, use has been made of maps, drawings, paintings, prints, photographs, and records such as rate books and Land Tax assessments, which taken together enable an over-all picture to be built up. Unfortunately, most of these sources are not as reliable as they might at first appear to be: maps and prints, although usually dated, may well convey quite erroneous information if not regarded with due circumspection.

In the case of maps, for instance, the marking of windmills gives a very sketchy guide to their actual existence. A map may show no mill when one undoubtedly existed at the time, due to either the mapmaker's unconcern for windmills or else the lack of space available for the mill symbol to be inserted in the appropriate position. The converse error of showing a mill where none existed was quite common among the earlier cartographers, as the attractive mill symbol served to fill up odd pieces of unoccupied space in a decorative fashion. In general, however, the study of maps can provide a basis for the location of mill sites. Windmills would have been significant landmarks in the sixteenth and seventeenth centuries, and are therefore important features of the maps of the time, although the pictorial type of mill symbol was conventionalized and was not intended to portray the mill as it actually appeared. Later on, a mill would not have been as noticeable a landmark in built-up areas, among factories, houses, and tall chimneys, and towards the end of the eighteenth century the mill symbol ceased to be used, being replaced by a hatched ground plan, in common with other buildings, especially on large-scale plans. It is interesting to note that the old-fashioned pictorial windmill symbol was retained by the cartographers of the Ordnance Survey, and appears in their maps to the present day.

However, a map may well show a mill which might not have existed at the time of publication, due to either the survey having been made some years prior to the date of publication, or the map being a later edition of a previous one without a revised survey having been made, or else to the map not being the result of an original survey at all, but rather a blatant copy of an earlier map by a different author, complete with mis-readings and spelling mistakes. The gradual mutation of place-names due to mis-reading can be traced on these copied maps. These are the main points to beware of as far as windmills are concerned; the study of errors in maps would warrant a survey to itself.

The dating of prints and drawings suffers from the same drawbacks as that of maps. A drawing may not necessarily be dated or titled by the artist at the time of execution, and may even be inscribed by a completely different hand, while the publication date of a print generally postdates that of the original drawing from which it was made by several years.

In general, the most accurate guides to dating mills are the records of monetary transactions relating to them, although even such documents as rate books show wide variation in the amount of information presented. Sometimes owner and tenant (either of whom might pay rates or taxes, and neither need reside at the mill, although presumably the tenant would be fairly near) are mentioned separately, sometimes not. Sometimes a separate rate is shown for a mill, with remarks as to its condition, but often there is no mention of the mill as such, the rate being included in a total for various properties under the name of the owner. Early rate books consist merely of lists of names, with no description of the properties concerned or details of their locations, and there are often large gaps in the chronological sequences due to loss or damage of the books over the centuries.

Although in the case of some Lambeth windmills no illustrations at all seem to be extant, an attempt has been made to present views of as many as possible, showing the positions of the mills in relation to their surroundings, close-up views showing their types and exterior features, and also detailed views of individual parts of the mechanism, or of the interior.

The development of the windmill in England

Before describing each site in detail, it would be useful to give here a brief outline of windmill development in this country. For a more complete and detailed study, the reader is referred to the several excellent books by windmill specialists mentioned in the bibliography. A glossary of technical terms is included at the end of this survey.

The earliest type of windmill was the post-mill, in which the wooden body of the mill, containing all the machinery and the stones, was balanced on the top of a large wooden vertical post about which it could be turned to face the wind. This post in turn was supported by a substructure known as the pyramid or trestle, consisting of two crossed horizontal timbers called cross-trees and four 'quarter-bars' which sloped from near the ends of the cross-trees up to the central post. This is the simplest type of mill, and can be seen illustrated in medieval manuscripts and early maps (*fig.69*).

The substructure of these post-mills generally rested directly on the ground; the timbers sometimes being embedded in a mound of earth to give added stability. Later on, the whole mill was raised above ground level by the insertion of four brick-built piers under the ends of the cross-trees. This arrangement removed the base timbers from the destructive effects of the damp soil, and the additional height thus given to the sails enabled greater driving power to be obtained. A further refinement was the complete enclosure of the substructure in a brick-built round-house, which protected the timbers from the elements and also provided useful sheltered storage space. It should be noted that the roof of the round-house in no way supported the weight of the mill, which was taken entirely by the post, as before. A downward extension of the weatherboarding of the mill body, known as the petticoat, served to exclude the weather from the gap between the body and the roundhouse roof.

From the base of the mill body at the rear, a long timber extended down to the ground level, and served to turn the mill to face the wind. This could be done either by the miller pushing against this 'tail-pole' with his shoulders, or by using ropes and pulleys attached to wooden posts fixed in the ground at intervals around the mill. Also at the back of the mill was the ladder giving access to the door in the mill body, and the tail-pole usually protruded through the steps of this ladder. Before turning the mill, a lever was operated to raise the ladder a short distance from the ground.

Over the centuries the post-mill developed from the simple single-roomed affair of early medieval times, to the grand multi-storeyed Suffolk post-mills of the nineteenth century, but the basic mechanical principle remained fundamentally the same.

During the seventeenth century another type of windmill made its appearance on the English scene. This was the smock-mill, so called because of its visual resemblance to the old type of agricultural labourer's clothing (*fig.9*). The smock-mill differed from the post-mill in that the body of the mill, containing the machinery, remained stationary, while only a cap at the top, carrying the sails, could be turned to face into the wind. The advantage of this arrangement was that a much larger mill could be built to contain more machinery and several pairs of stones, without the difficulties inherent in constructing a trestle strong enough to take the greatly increased weight. Like the post-mill, the smock-mill was built of wood, but the rectangular plan was discarded in favour of an octagonal, or sometimes hexagonal, form, to allow clearance for the sails as the cap turned round the mill body. For increased stability, the sides of the mill sloped inwards to the top, giving the mill its characteristic 'batter'. Most smock-mills were built on top of a low brick wall to protect the lowest timbers from the damp, but later this developed into a whole storey built of brick, with a stage running round the outside to enable the sails to be reached for adjustment. Instead of turning about a central pivot, the cap rested on a track fixed to the top of the mill walls, and slid round on this. A later refinement was the insertion of two sets of rollers, to centre the cap and to reduce the friction in turning.

One of the most serious disadvantages of both the post and smock-mills was their timber construction, and hence vulnerability to the vagaries of the weather. To overcome this, smock-mills were eventually built entirely of brick, still sloping inwards towards the top, but circular in plan, with only the cap being made of wood. This type is called the tower-mill (*fig.49*), and accounts for the majority of windmills surviving today, most of their timber brethren having long since succumbed to the elements. Various hybrid mills, such as the hollow post-mill, in which the machinery was in the lower part, and the composite mill, consisting of half post-mill and half tower-mill, have made their appearances from time to time, together with curiosities such as the horizontal mill which stood at Battersea, but the three main types have remained the post, smock, and tower-mill.

In parallel with the structural development of the windmill ran the development and improvement of the various parts of the mechanism, the most immediately obvious being the sails. The earliest type of sails, known as common sails, consisted simply of sheets of strong canvas stretched over wooden frameworks. The canvas could be adjusted by means of ropes, so that varying areas could be exposed to the wind. This procedure involved using formalized positions of the canvas, called by such names as first reef, sword point, and dagger point. At first, the rectangular frames of the sails were given the same twist all along their length, but it was later discovered that better results could be obtained by varying the twist from a maximum of 18° to 7° at the tip. This was confirmed by the experiments of the distinguished engineer John Smeaton, who read a paper before the Royal Society on the subject in 1759[1]. The sail frameworks were built up around central timbers called 'whips', which were fixed to other timbers ('stocks') mortised at right angles through the windshaft. Later a cast-iron canister, or 'poll-end' was used to replace the mortised joint, and sometimes the stock was dispensed with altogether, and the whip, being made larger and stronger, bolted directly onto an iron cross fixed to the end of the windshaft. In early mills the sail extended on both sides of the whip, but later the sail was fixed only to the trailing, or driving, side, with a wooden board on the leading edge to give a streamlining effect. A great disadvantage of this type of sail was that the adjustment had to be made from the ground while the sails were at rest. (The sails had to be very finely balanced to enable each one to be turned to ground level.) It was thus very difficult to reduce the area of sail when a strong wind sprang up, as application of the brake ran the risk of setting the mill on fire due to friction. In cold weather the canvas would freeze stiff, rendering any adjustment virtually impossible. To overcome these difficulties, a type of sail consisting of separate divisions holding shutters instead of sheets of unbroken canvas was invented, known as the spring sail.

This type of sail was invented by Andrew Meikle in 1772. The shutters were made of wood or of canvas stretched on frames of iron or wood. Each shutter was hinged, and they were all connected to a wooden shutter bar which ran down the length of the sail, parallel to the whip. On each sail an adjustable spring controlled the shutter bar, and this could be set so that at a given wind pressure the force of the spring would be overcome, and the shutters would rotate, allowing the wind to pass through the sails. This was known as 'spilling the wind'. The main disadvantage of this type of sail was that it could not be constructed with the correct amount of twist, and hence the power obtainable from it was not as much as that given by the common sail. As a consequence, windmills were often fitted with a pair of spring sails and a pair of common sails, in order to give power together with a degree of regulation (as at Ashby's mill, Brixton) (*see fig.48*).

Although the adjustment process was much simpler with the spring sail, the mill still had to be brought to a halt in order to make the adjustment. In 1789 Captain Stephen Hooper invented a new type of sail which enabled adjustments of the sail area to be made while the mill was still in motion. Known as the roller-blind sail, it used small roller-blinds in place of shutters in each section of the sail, the blinds being attached to each other by means of canvas strips called listings. A hole was drilled through the centre of the windshaft, and in this was fitted a rod, called the striking rod, which was connected by means of a lever system (the 'spider') to the air poles which enabled the blinds to be adjusted simultaneously. By arranging a gear system connected to the striking rod, the amount of extension of the roller-blinds could be controlled while the sails

2. A millstone at Ashby's mill, Brixton, showing the characteristic system of surface grooves, or 'furrows'.

were actually turning, and could be set to a given amount by means of an endless chain on which varying weights were suspended. This was a great advantage, and overcame the risk of fire due to hurried application of the brake in gusty weather. However, the blinds and listings needed constant replacement as they were quickly attacked by damp.

A final refinement of these various devices was the patent sail, invented by William Cubitt in 1807. This combined the shutters of the spring sail with the striking rod mechanism of the roller-blind sail. Sometimes the striking gear was controlled by a centrifugal governor which allowed the sails to spill the wind when the speed of rotation exceeded a given amount, but generally the suspended weight system was used.

Most windmills had four sails, but occasionally experiments were made with five, six, or even eight-sailed mills. Although these were smoother running and gave more power than the normal type, they were more costly to maintain and were not generally adopted. As far as can be ascertained, all the mills in the Lambeth area were of the four-sailed variety.

Early smock and tower-mills were provided with long tail-poles reaching from the cap to the group or the stage, to enable the sails to be turned to face into the wind. This arrangement was superseded by a rack and gear system which could be operated from the ground by means of an endless chain mechanism. By pulling on the chain, a cog was made to act on a rack fitted to the top of the mill body, and thus the cap was turned round.

An improvement on this system was the device known as the fantail, or 'fly', invented by Edmund Lee in 1745. This consists of a small six or eight-bladed windvane fixed at right angles to the main sails, either at the rear of the cap, or, in the case of a post-mill, to a staging built onto the external stairs. When the sails are facing directly into the wind there is no pressure on the fantail, which remains stationary, but if the wind veers to one side it immediately causes the fantail to revolve. By means of a set of gears the movement of the fantail is made to act on a rack in the same way as the endless chain arrangement described above, and the cap is turned once more into the eye of the wind. In this way the sails are kept facing into the wind quite automatically, thus relieving the miller of the continual and irksome task of winding the mill at each change of wind direction. The possibility of the wind catching the sails from behind and overturning the mill was thus virtually eliminated.

Internally, the basic principles of windmill mechanisms have remained unchanged since medieval times. The drive from the windshaft is taken through a set of gear-wheels to one pair or more of millstones which are arranged horizontally, the upper, or 'runner', stone rotating and the lower ('nether') remaining stationary. Corn is fed from a hopper to a hole in the centre of the runner stone, whence it passes between the stones and is ground into flour and bran which are collected underneath. In early mills all parts of the mechanism, including the gear-wheels, were made of wood, but the use of cast-iron later became widespread, with a consequent increase in efficiency and reliability, although involving a considerable increase in weight. Several kinds of millstone were in common use; French, or Burr, stones of quartz were used for grinding flour, Peak stones of millstone grit quarried in the Peak District of Derbyshire were used for animal feed, and Cullin or Blue stones from Cologne were used mainly for grinding barley and oats. A series of grooves, or 'furrows', was cut into the surface of the stones, and the maintenance of these, i.e. 'dressing' the stones, was a regular part of the millwright's work (*fig.2*).

To this basic mechanism, various supplementary devices were gradually added in order to increase the efficiency of the mill and ease the load of work on the miller. The first and most important requirement was a brake to enable the sails to be brought to a halt. This generally acted on the periphery of the first gearwheel on the windshaft, thus giving it the name of brakewheel. Other devices included a vibrating feed trough which ensured an even supply of grain to the stones, a warning bell to indicate when the hopper was nearly empty, and a sack hoist driven from the windshaft to enable the sacks of grain to be lifted to the top of the mill before grinding. In later windmills machinery was installed to dress the flour after it had been ground. The most common types were the bolter and wire machines, which separated the flour from the bran immediately after it had passed between the stones, and which were driven from the main drive shaft.

However, in spite of all these ingenious mechanical devices, the windmill was no match for the mass-production techniques of the Industrial Revolution, and many mills simply went out of business in the face of intense competition. In Lambeth only one windmill survived the tremendous industrial expansion which swept over the area during the nineteenth century, and only by piecing together the various fragments of evidence is it possible to obtain an idea of what the mills were like, what they were used for, and of the people who operated them.

Windmills in Lambeth

The earliest evidence of a windmill existing in the area now covered by the borough of Lambeth dates from the end of the sixteenth century, although there were doubtless mills here during the Middle Ages, as there were in the majority of rural districts in England. From that time to the present there has always been a windmill of some kind in existence, and some twelve sites have been recorded altogether, although the maximum number standing at any one time was five (at the beginning of the nineteenth century) (*fig.1*).

Windmills are usually associated with corn milling, but in Lambeth the spread of industrialization caused the power of the wind to be set to others tasks, and we find mills driving mechanical saws for timber yards or used in the manufacture of starch and mustard, and also for crushing materials used in the production of chinaware.

In addition to the mills themselves and the millers who worked them there were associated trades, such as mechanics and millwrights, who appear from time to time in the trade directories (although they would have also dealt with other types of mill besides windmills). John Edwards, millwright and pump maker, is listed in Kent's Directory[2] of 1812 as being at Vine Street (now covered by the site of Waterloo Station), and the firm of Parsons & Woods, millwrights, had premises 'near King's Arms Stairs, Lambeth' (i.e. on the river front, about halfway between the present County Hall and the Royal Festival Hall). By a lucky chance, the trade card of this firm has survived (*fig.3*), and besides being beautifully engraved, it tells a lot about the nature and versatility of their work:

> Parsons and Woods, Millwrights and Engineers, near King's Arms Stairs, Lambeth, Surry. Make & Repair all sorts of Millwork, Engines, & Pumps for Raising Water to the Greatest Perfection. Also Make & Sell French Stones, French Burrs, Plaister of Paris, & all other Sorts of Millstones, at the Lowest Prices. N.B. Gentlemen, Millers or others by sending their Orders shall be as Well Serv'd as if they came themselves, We being Determin'd to Deal upon Honour.

The card is well embellished with decoration, including pictures of a tower-mill with a projecting gable covering the winding gear (a typical feature of many Lambeth windmills), a double-cylindered water pump, and craftsmen engaged in dressing millstones[3].

As the built-up area expanded at an extraordinary rate during the nineteenth century, the windmills were superseded by more sophisticated and efficient machinery, and by mid-century they were all gone except one, which lingers on to the present day although no longer capable of working. The following survey attempts to describe each mill, to locate its site, and to give as much information about it as can be obtained from pictorial and documentary sources.

3. Trade card of Parsons & Woods, a Lambeth firm of millwrights. From the collection of Sir Ambrose Heal.

Site 1: North Lambeth

This mill stood on the south bank of the Thames on the bend opposite the Savoy; the site now being occupied by the Royal Festival Hall arts complex.[4] It is marked on no less than fifty-two different maps, commencing in 1658 and running well into the eighteenth century, but the documentation is not so thorough as it may appear, since many of these maps are manifestly copies of earlier maps by different authors, while others are later editions of earlier versions which have been reprinted without alteration.

The origins of the mill are described in the fifth volume of Cox and Hall's *Magna Britannia* (1730) in the following terms:

> In this Parish was erected a Saw-Mill during the Usurpation of *Oliver Cromwell*, which was then such a Rarity in *England* (being perhaps the first of the Kind) that *Cromwell*, tho' Protector, often went to see it, and was so much pleased with the Contrivance, that nonwithstanding the Clamours of several Artificers about it, he got it confirmed by Act of Parliament. It stood in *Belisderes* Gardens.[B28]

The date and function of the mill are confirmed by its first appearance on a map, that of Newcourt dated 1658.[5] Here it is shown as a post-mill (a conventional symbol used by cartographers) and although not marked in words as a sawmill, planks of timbers are shown pictorially lying by the side of the mill. The form of the mill follows the usual method of representation; a front view with no windows and a straight ridge roof. An interesting point is that an open pyramid is shown below the mill, whereas the symbols used for the windmills in north London show the substructures as being buried in small mounds. These symbols, however, were doubtless not intended to depict the mills concerned, but merely to serve as a conventional indication of their sites. Saxton's map of *c.*1690[B34] has 'Sawmill' written against the mill symbol, and a long strip of the Surrey shore towards Lambeth Palace is marked as 'All wood yards'.

No mention of the Act of Parliament referred to by Cox and Hall can be traced amongst the legislation passed during the time of the Commonwealth, but the opposition of the 'Artificers' rings true, as there are other cases of workmen acting to protect their interests against the onslaught of the new machinery. According to Heckethorn in his *London Memories*,[B26] a sawmill was built by a Dutchman at Limehouse in 1663, but the project had to be abandoned owing to the opposition of the sawyers, and again in 1768 another sawmill was built at Limehouse which had to be rebuilt after having been pulled down by a mob. As to whether the Lambeth mill was 'the first of its Kind', Heckethorn states that the Limehouse mill was the first sawmill built in London, but as he gives a date of 1663, it was clearly pre-dated by the one at Lambeth.

penetrated to London. Such mills were generally built close to a river due to the cheapness and ease of transporting timber by water, and differed from the more usual types of windmill in that the entire structure, including the projecting sheds fixed to each

4. Detail from an engraving by Pieter Schenck, showing the North Lambeth sawmill.

5. Detail from a view of Lambeth, Vauxhall, Millbank and Westminster, by Francis Place.

6. Part of another view by Francis Place, in which the sawmill appears as a prominent feature of the Thames-side landscape.

The exact type and construction of the mill can be ascertained from two contemporary illustrations, the first an engraving by Pieter Schenck (*fig.4*),[6] and the other a pen-and-ink sketch of Lambeth, Vauxhall, Millbank and Westminster made by Francis Place in about 1683 (*fig.5*).[7] Both of these illustrations, which otherwise seem to be accurate in their general topographical details, show the mill standing among low buildings and piles of timber, a few yards from the water's edge, and in both cases it is shown as apparently having a square body, with low sheds attached to either side. From these details it is evident that the mill was a sawmill of the type known in Holland as a 'paltrok' mill, so-called because of its visual resemblance to the dress of the women of the Zaan district. This type of windmill was invented by Cornelis Corneliszoon, of Uitgeest, and patented by him in 1592, and so the middle of the seventeenth century seems a reasonable date for the idea to have side, could be turned to face into the wind.[B5] This accounts for the characteristically square shape of the body, which was always in the same position relative to the sails and hence did not need to be round or octagonal as was necessary in the case of a tower or smock-mill to allow a clear passage to the turning sails. Another view by Francis Place, entitled 'Westminster from the Middle Temple Garden'[8] shows the mill in the distance as a prominent feature of the Thames-side landscape (*fig.6*).

Further evidence supporting the date of construction is provided by pictorial records. The mill is not shown in Hollar's view of London from Bankside made in 1647, but it does appear in Jonas Moore's 'Prospect of London' (1662), and Hollar's 'Prospect of London and Westminster, taken from Lambeth' of 1674.[9] In both of these engravings it is shown as a post-mill with a straight-pitched roof; again, the

7. Detail from John Kip's view of 1710, taken from Buckingham House, showing part of the south bank of the Thames, now the site of the Royal Festival Hall.

stylized representation of windmills as found in maps of the period. Moore gives the mill two windows in the back, and shows the mill in St George's Fields in a similar way.

The paragraph from *Magna Britannia* implies that at the time of writing the mill no longer stood, although it was certainly still standing as late as November 1710, when an insurance policy was issued by the Hand-in-Hand company

> on a brick house situate at the North End of the West Side of Lime Tree Court on the North Side of the Narrow Wall between Cuppers Bridge and the Windmill in the parish of Lambeth St. Mary fronting north to a timber yard and south on a Garden . . . [10]

In the same year John Kip's 'Prospect of the City of London, Westminster and St. James's Park, taken from Buckingham House' was published,[11] showing the mill standing close to the water's edge, between two low buildings, presumably warehouses or bargehouses (*fig.*7). Narrow Wall is now Belvedere Road, which runs past the entrance to the Royal Festival Hall, and whose name commemorates the Belvedere Gardens mentioned by Cox and Hall (although mis-spelt by them).

The mill continued to be marked on maps well into the second half of the eighteenth century, but most of the later ones are copies of earlier maps,[B33] and it is probable that the windmill disappeared about 1750, if not before. The last map on which it appears is that published by Sayer in 1775,[12] which is a later edition of C. Dicey and Co.'s map of 1765, and several other maps of the 1760's do not show the mill.

8. The North Lambeth sawmill as seen across the marshes from Lambeth Palace: detail from a view by W. Hollar.

The words 'Saw Mill' appear quite frequently on the maps before this date, both with and without a mill symbol, and the misprint in Roades' map[13] of 1731 which gives 'San Mill' is faithfully copied from an earlier map which perpetrated the original error.

Practically nothing is known about the owners or tenants of the mill, except that two maps give owners for the land on which it stood. Morgan's map[14] of 1677 shows the mill standing in a plot of land marked as 'Alderman Smith's', while Jefferys' map[15] of 1732 shows the same plot of land as belonging to a 'Mr. Mare'.

Heckethorn, in his *London Memories* [B26] claims that 'in the winding street then called Narrow Wall, now Belvedere Road, there was a perfect group of picturesque old wooden windmills', but no evidence has been found to indicate that more than one mill ever stood on the site.

Site 2: The Cut (formerly New Cut)

Turning inland from the river bank, the land which lies between what is now Stamford Street and The Cut remained undeveloped until the early nineteenth century, and was known as Wild Marsh.[B36] It was dotted with occasional clusters of buildings and was crossed by a few paths such as Glovers Halfpenny Hatch, whose name indicates the toll payable by persons wishing to pass across the marsh. The north side of New Cut was occupied by a few buildings, and it was amongst these that the windmill stood.[16]

It is first mentioned in a survey of the Manor of Kennington, to which the land belonged, made by Middleton and Hodskinson in 1785–6.[17] The proprietor of the whole piece of land is given as Jane Curzon, and the property is described as

> All those two Brick Dwelling Houses, Brewhouses, Starch Houses, Windmill, Stables, Carthouses, Pigsties, Sheds, Yards and Gardens, Situate on the East Side of Green Lane (*i.e. what is now Cornwall Road*).

The survey further states that the property was 'formerly in the Occupation of John Robbins and lately of John Wigley, but now of Messrs Lawrence & Wells.'

The first edition of Horwood's large scale map[18] dated 1794 shows the mill as a round hatched circle, with no explanation, but the second edition[19] of 1807 adds the word 'windmill' to the circle to confirm the representation. Its approximate position from this map was 135 yards from the junction of New Cut and the Waterloo Road, and 40 yards north of the Cut itself, thus locating it in what was later to become Little Windmill Street (now Cons Street). The existence of the mill is commemorated in the name Windmill Walk, a turning off the Cut, and the Windmill public house, which stands on the corner at the junction of the two streets.

On Laurie and Whittle's map of 1800[20] the mill is shown by means of a tower mill type of symbol, and a watercolour painting by William Capon entitled 'View of London from a Gentleman's seat in Lambeth Marsh' dated 1804 shows it to be an eight-sided smock-mill with four sails and a stage supported by struts; the cap being apparently semi-spherical in shape.[21] The area around the mill appears to be fairly well built up, even at this date. More detail may be seen in an excellent watercolour by H. de Cort,[22] dated August 1804, and entitled 'A mill in Lambeth Marsh road on the left leaving Westminster Bridge' (*fig.9*). It is shown as an octagonal smock-mill of three storeys, built on a three-storeyed wooden base. A stage runs round the top of the base and is supported by struts where its circular shape diverges from the rectangular shape of the base. The mill has four common sails, and an almost ogee-shaped cap with a small gable at the rear to house the winding mechanism, from which chains or ropes hang down to the level of the stage.

9. 'A mill in Lambeth Marsh road on the left leaving Westminster Bridge': a watercolour by H. de Cort, dated August 1804.

A 'View of Southwark, with part of St. George's Fields, from Beaufort Buildings, Strand' by Malton, published by W. Richardson in June 1807 shows the mill behind some trees and a row of houses, as seen from the north bank of the Thames (*fig.10*). It is shown as a smock-mill, but no further details are represented. The houses in the background of this view have more the aspect of an Italian hill town than of the Lambeth marshes, but it is a useful indication that the mill was standing at the time. Walford's date of 1770 for this illustration, given in *Old and New London*,[B30] is obviously incorrect, as the view includes the square shot tower which was not built until *c.*1789, and his attribution of the authorship to Nugent is quite erroneous as this artist's name does not appear anywhere on the original print.[23]

The first mention of the mill in the Lambeth Rate Books[24] occurs in May 1805, when the entry 'John Rusby for the Mill' appears. The annual value of the premises is given as £32. By September the following year, however, the tenancy had passed to the firm of Woodcock & Keap, who remained for just over a year. In December 1807 the occupier is given as William Southall, who appears throughout the Rate Books for the next three years, until he relinquished his tenancy to Joseph Orchard sometime between May 1810 and June 1811. Orchard remained at the mill for several years, his landlord's name being given as Joshua Lee, but in May 1816 the previous entry 'Joseph Orchard for House and Mill' becomes simply 'occupier', indicating that a change had taken place. Unfortunately, there is a gap in the sequence of Rate Books until the year 1824, when the mill is no longer mentioned, but it seems certain that Orchard left the mill early in 1816.

The first edition of the one-inch Ordnance Survey map published in 1816 shows the mill,[25] as does Wallis' map of 1821, but the latter is probably a copy of Laurie and Whittle's map of 1809 and shows the mill in exactly the same way as a circle, marked 'Mill', against an oblong building.[26] Although the third edition of Horwood's map published in 1813 shows the mill in the same way as the second edition, the fourth edition of 1819 shows the site as having been built up during the intervening years, and the windmill is replaced by a street marked as 'Upper Windmill Street'. It is clear from this that the mill was demolished between 1816 and 1819 to make way for houses which were being erected by the property developer John Roupell. According to a newspaper cutting formerly in the possession of Lancaster Burne[27] the mill was pulled down in 1840, but there is no evidence to substantiate this assertion, and there are certainly no maps dating from later than 1816, apart from the one by Wallis mentioned above, which show the mill. R. Thurston Hopkins, in his *Old Watermills and Windmills*[B13] describes Windmill Street, New Cut, as marking the site of 'a group of picturesque wooden mills', but once again no more than one mill on this particular site is ever mentioned in any of the documentary evidence.

The exact date of erection of the mill is uncertain, as no maps cover the area in sufficient detail in the latter half of the eighteenth century, but Walford gives a date of 1750 for an engraving which shows two windmills, both seemingly post-mills on high round-houses, one of which occupies the same position as the mill shown in Malton's view of 1807. Whether this represents the same mill, or whether the smock-mill replaced an earlier post-mill on the same site is open to speculation, and the exact whereabouts and nature of the second mill shown in this print have proved to be beyond verification (*fig.12*).

10. Detail from Malton's view of 1807, showing the New Cut windmill and the square shot tower.

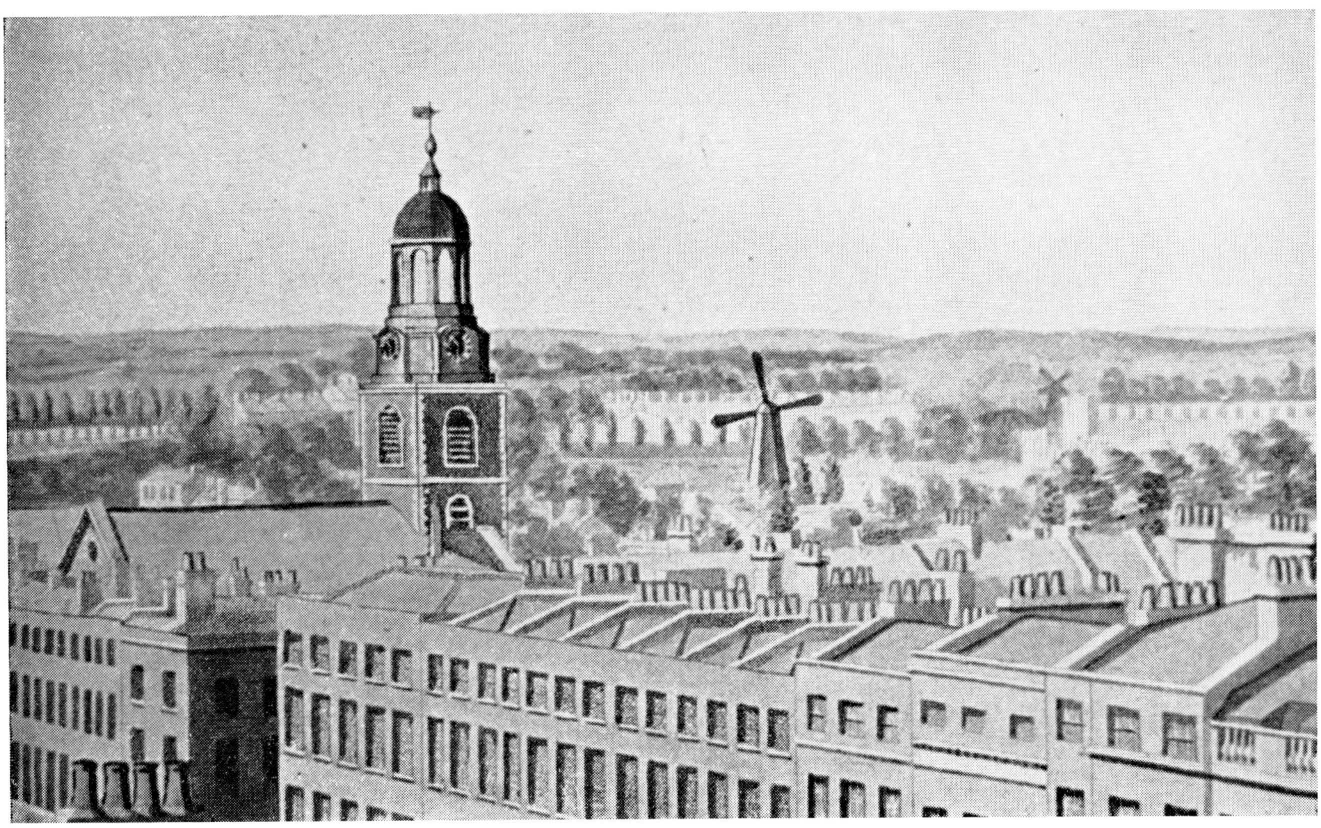

11. The New Cut smock-mill as seen over the rooftops from Southwark: detail from an early nineteenth-century print.

12. Detail from an eighteenth-century engraving, showing two post-mills in the vicinity of New Cut.

Site 3: Juxon Street (formerly Mill Street)

During the eighteenth century three windmills were erected in the vicinity of Lambeth Walk, then known as Three Coney Walk (*fig.13*). The most northerly was situated in a field to the west of the Walk and north of Paradise Row, and was approached by means of a lane which later became known as Mill Street (now Juxon Street)[28].

The first evidence of its existence is a tinted drawing by Bernard Lens,[29] dated 1735, entitled 'A view of the Palace of Lambeth', which shows the mill standing behind the Palace, as seen from the river. Not much detail can be ascertained from this drawing, but it is apparently shown as a tower-mill with four sails. A similar view of the Palace dating from ten years previously does not show the mill,[30] and in the absence of any other evidence a construction date of about 1730 must be assumed. Another view of Lambeth Palace, by Samuel and Nathaniel Buck,[31] dated 25 March 1737, also shows the mill. It appears as a tall tower-mill in this engraving, with something of an elongated sugar-loaf shape. Its first appearance on a map occurs in that of Rocque,[32] surveyed in 1741–5, where it is shown by means of the conventional post-mill symbol, and from this date onwards it appears in the majority of printed maps, although with some curious omissions. For instance, it is not marked on Rocque's map of 1773, nor on Lindley and Crosley's of 1789, although it was undoubtedly standing at the time.[33]

A fairly complete list of the millers who were associated with the mill is provided by the Lambeth Rate Books[34] and the assessments for Land Tax,[35] both of which give details of tenants and rateable values of properties, although they do not give a continuous coverage over the whole period of the mill's existence. The first name connected with the mill is mentioned in the Rate Book for May 1770, where Thomas Corner is charged fifteen shillings on a rateable value of £30. The same entry appears in the following two years, and in 1773 the nature of the property being rated is revealed by the entry 'Thomas Corner for ye Mill'. This entry is also repeated in the remaining books for 1773 and 1774, but in December 1775 we find 'John Stephens for ye Mill', which is repeated in 1776. Here there is a gap in the sequence of Rate Books, but further information may be had from the Land Tax assessments. The first mention of the mill appears in the year 1780, where the owner is given as William Chilwell, Esq., and the occupier as John Marler; the taxable value being £25. According to the assessments, Marler continued in the tenancy until 1784, when the mill was taken over by Robert Machell & Co., although the owner remained

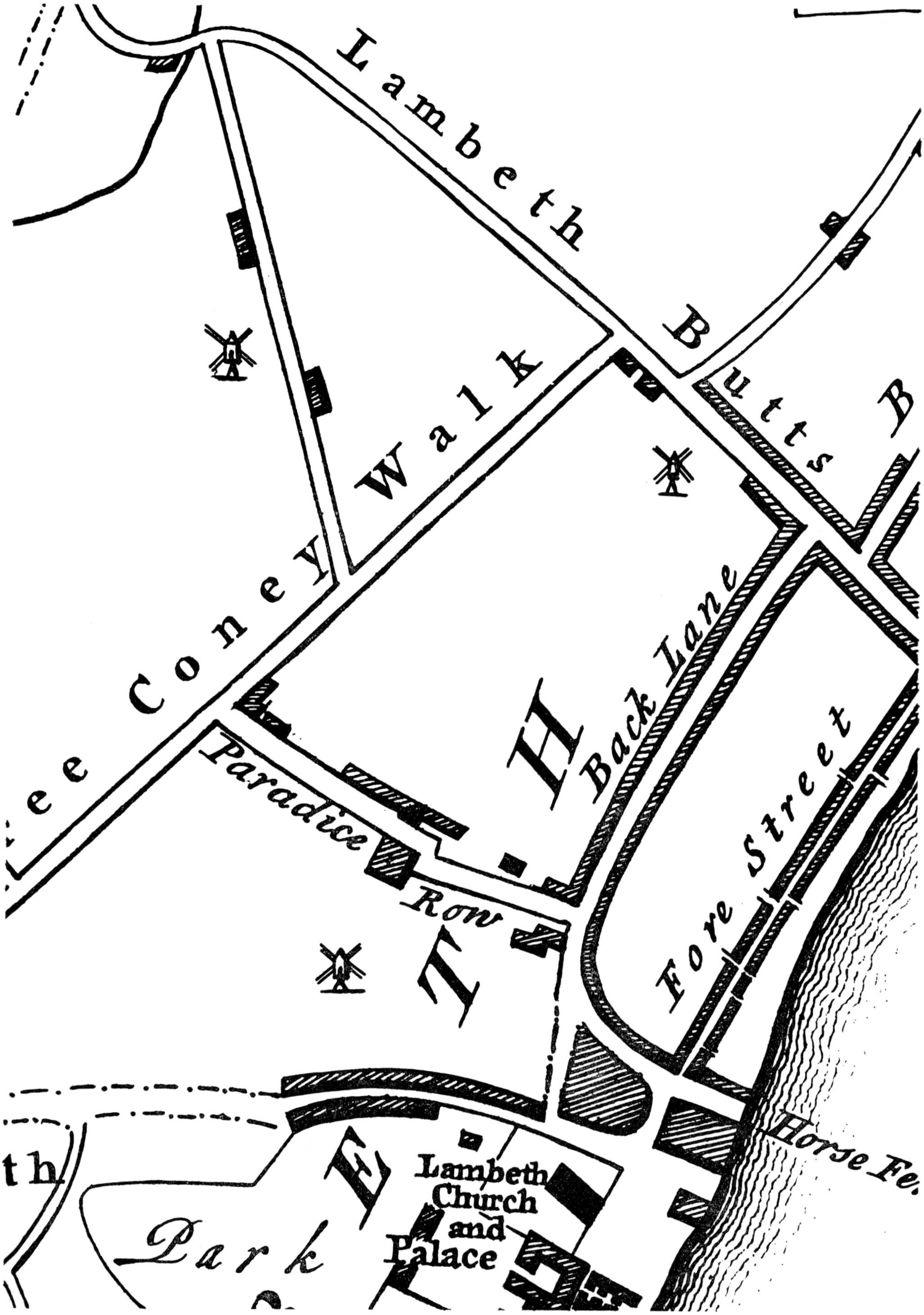

13. Detail from 'A Map of the New Roads from Westminster Bridge' published in 1753, showing the three windmills near Lambeth (Three Coney) Walk.

14. 'Lambeth Terrace, near the Palace': a watercolour of about 1810.

the same until 1786 when Jonathan Chilwell is listed, presumably the son of William. During the last few months of 1788 Robert Machell died, and the rate made in December is assessed to 'The Executors of the late Robert Machell'. Following Machell's death the ownership was transferred from Jonathan Chilwell to Samuel Swabey, Esq., but the mill remained empty for nearly two years until another tenant was found.

In September 1790 John Clark took over the mill and in December 1792 he was joined by James Clark. Their joint tenancy was short-lived, however, as the mill is again listed as being empty in August 1793. This time it took longer to find a suitable occupant, and it was not until March 1796 that Richard Herringshaw took up the tenancy, by which time the ownership had passed to Sir Joseph Mawby. Herringshaw remained until 1802, when there were two very brief tenancies, of one 'Stevens', and John Wiple, and then in September 1803 Susan and Robert Tabrum took over. These two remained until September 1807, and the mill then once again lay unoccupied until January 1809, when John Alcock moved in. However, he in his turn soon left, and by December the same year we find John Fleet listed as tenant. This miller was to come to an unfortunate end to his tenancy; he is listed in both sets of account books for 1810 and 1811, but in 1812 the windmill is not mentioned at all. The explanation can be found in the issue of the *London Gazette* dated Saturday, 3 August 1811, which states that a Commission of Bankrupt was 'awarded and issued against John Fleet of Mill-Street Lambeth, in the County of Surrey, Miller and Coal-Merchant' (possibly a mis-hearing of 'corn-merchant') and that he was 'required to make a full disclosure of his estates on the 10th and 17th August inst. and 14th September next'.[36]

Several illustrations dating from about this time give a considerable amount of information regarding the appearance of the mill and its construction. A watercolour of about 1810, inscribed 'Lambeth Terrace near the Palace' (*fig.14*), shows a tall, ten-sided smock-mill with a stage and four common sails,[37]

15. A print dated 1814 showing the tall smock-mill in Mill Street (site 3).

16. 'Old mill, Lambeth': a watercolour by H. Pyall, dated 1820.

while a print dated 1814 shows much more detail (*fig.15*), including a fantail which is apparently fixed onto a gable housing the previous hand-winding mechanism.[38] From this illustration it appears that the mill contained seven storeys; the lower three being circular and stone built, and the upper ten-sided portion being of wood, with oblong windows on each floor. There are four common sails with furled canvas, and a semi-spherical cap. A well-executed watercolour by H. Pyall,[39] dated 1820, confirms these details, and a petticoat seems to have been added around the lower edge of the cap and gable, to exclude the effects of the weather from the winding mechanism (*fig.16*). A tall chimney rises beside the mill; possibly the chimney of an associated bakehouse. This painting fixes the location of the mill in relation to the tower of Lambeth Church, which appears in the background (omitted from fig.16), and the rapid spread of building throughout the area is apparent from the fact that the mill seems to be hemmed in by houses on all sides. Only eight years previously, in 1812, a map by A.W.E. & S. Driver showed the area to the south of Mill Street as open ground described as 'Potters Fields'.[40] Another illustration dating from 1820 is a view of London and Westminster by T. Nugent,[41] which shows both this mill and the one to be described in the next chapter as tall six-sided smocks with stages; the sails being so badly drawn that they appear to be hanging in mid-air several feet above the top of the mill body! (*fig.18*). A much more accurate and informative illustration is a sepia drawing dating from 1828, by the architect John Buckler (*fig.17*). In addition to the details shown in the other illustrations, two doors can be seen, giving access to the stage from the mill body, and under the stage there is a row of circular windows. The construction of the fanstage can be clearly seen in this view.[42]

Following the bankruptcy of John Fleet, the operation of the mill was taken over in February 1813 by B. Holden, whose tenancy lasted until 1820. For three years following this date, the Land Tax assessments merely list 'The Mill', at a value of £70, but as no tenant's name is given it is not clear whether it was actually in use during this period. In 1824 'James Nokes for Mill' makes an appearance, and this is borne out by Pigot & Co.'s London and Provincial New Commercial Directory for 1823–4, which gives in the Millers section: 'Noakes, J. Windmill-St. Lambeth'.[43] This Noakes was probably a refugee from the saga of the Westow Hill windmill, which will be recounted in a later chapter. His namesake abandoned the Westow Hill mill in 1821, and so the appearance of the name at Mill Street in 1823 tends to suggest that this was one and the same person. However, his new tenancy did not last for very long either, as the mill is listed as being empty in 1825 and also in the two following years.

On 6 August 1827 the following advertisement appeared in the *Reading Mercury*:

> To Mealmen, Bakers, etc. To be disposed of by Private Contract, the lease of Lambeth Mills, Mill Street, Lambeth, consisting of a powerful Windmill, steam engine with six pairs of French stones, dressing machinery, smut machine, with tackling complete; also a bakehouse with a 10-bushel oven, with extensive premises, the whole recently put in perfect repair, also a neat cottage adjoining the same. Apply on the premises, or at 148 Tottenham Court Road, London.[44]

This seems to have had some result as, after a gap in 1827, the mill is again mentioned in the Land Tax assessments for 1828, where the 'occupier Windmill' is taxed on a value of £80, and the following year further detail is provided by the entry 'Kingsford Windmill'. Here the Land Tax books cease, but there are two further entries in the Rate Books, one for October 1839, where 'Ashford - Mill' is rated on a value of £90, and in September the following year 'Davidson' is rated for 'Mill and tenements' at a value of £120. The Post Office London Directory for 1843 includes in its list of millers the entry 'Davison William, Mill St. Lambeth'.[45] The Census of 1841 mentions several millers in the area, but whether they worked in this mill is not certain.[46] Thomas Vine (aged 40) and family are listed in Mill Street, David

17. 'Windmill in Mill Street, Lambeth': a sepia drawing by John Buckler, dated 1828.

Knight (25) and his wife in Windmill Street (which was approximately on the site of the present-day Sail Street), and John Athill (30) in Lambeth Walk.

There are a few views in which the mill appears in its later years. Leigh's Panorama of the Thames (*c.*1830) shows it to the right of Lambeth Palace as seen from the river, and as the rest of this panorama seems to be quite accurate it is possible to make an estimate of the height of the mill in comparison with other buildings.[47] It appears to be about two-thirds the height of Lambeth Church tower (100 ft), and allowing for the effects of perspective, this supports Farries and Mason's estimate of 75 ft to the top of the cap. A lithograph entitled 'Westminster from Waterloo Bridge' of about 1840 shows the mill above the roofs, between a tall chimney and a warehouse, and it is interesting to note that the representation of the mill bears a striking resemblance to its depiction in the 1737 engraving of Lambeth Palace mentioned previously.[48] One further view exists, dated 1841, in which the mill and Lambeth Palace appear in the background of a boat-race scene.[49] The perspective is curiously distorted and the mill is drawn as a tall, straight-sided building with a straight pitched roof and with four sails. Not much can be gleaned from this picture, but it serves to indicate that the windmill was still standing at this time.

Maps continued to mark the mill for several years, and the last one on which it appears, that of Cruchley dated 1847,[50] also shows the route of the railway line extension which was under construction from Nine Elms to York Road (Waterloo) passing close to the west of the mill site, and it was undoubtedly this which caused its demise. The extension to the railway was officially declared open on 11 July 1848.[51] No further documents or maps make any mention of the mill, and the Census of 1851 records Z. Blavean, Stove Maker, as carrying on his trade in Mill Yard, Mill Street, the mill itself apparently having been demolished.[52]

Site 4: Whitgift Street

Fifty yards north of Broad Street (now Black Prince Road) stood a windmill which was approached by means of a pathway from the junction of Broad Street and Lambeth Butts, its site approximately corresponding with the point at which the railway arch now crosses Whitgift Street.[53]

Like the one in Mill Street, its first appearance on a map occurs on that of Rocque, surveyed in 1741–5,[54] where it is shown by means of the usual post-mill type of symbol. Several subsequent maps indicate the mill in a similar fashion, and on the first edition of Horwood's large scale map dated 1794[55] it is shown as a round building standing in a yard marked as 'Messrs. Stonard & Watson's Starch Manufactory'. The firm of Stonard, Watson & Weymouth is mentioned in the Lambeth Rate Books for this period, being rated for premises previously under the name of Jonathan Stonard, Esq., but the windmill itself is not mentioned as such, and it has not been possible to ascertain as to whether it was actually involved in the starch manufacturing process or whether it was in use separately as a flour mill. In August 1804 the factory premises are listed as being unoccupied, and in May 1805 a new tenant, William Holden, is rated for the windmill itself, on an annual value of £50. Again, it is not clear whether the mill was in use separately or in conjunction with the associated factory premises. The Lambeth Enclosure map of 1806 shows the mill as octagonal in plan,[56] and marks the whole premises as 'Mustard Manufactory', as does the second edition of Horwood's map, published in 1807.[57] Between September 1808 and January 1809 Holden left the mill, and may well have been subsequently connected with the firm of Holden & Co. which took over the Mill Street windmill in 1813 following the bankruptcy of John Fleet, described in the previous chapter.

In January 1809 the firm of Weymouth Carpenter & Rutter was rated for both the windmill and the factory premises, which the Land Tax assessments variously describe as 'Manufactory', 'Steam Mill', or 'Starch Mill'. The explanation of the variation in the descriptions of the premises is probably that production of both starch and mustard was carried on at the same time in different parts of the factory, utilizing the windmill in the manufacturing processes according to the particular work in hand.

In 1820 a change occurs and the name of Joseph Orchard appears as a separate tenant of the windmill, although Weymouth, Carpenter & Rutter continue to be assessed for the remainder of the premises. It will be remembered that a Joseph Orchard was in

18. Detail from a view of London and Westminster by T. Nugent, dated 1820, showing in a rather crude fashion the mills at sites 3 and 4.

charge of the New Cut mill until 1816, and it is likely that this was the same person. Most of the entries in the Land Tax and Rate Books describe the premises simply as 'Windmill', or even 'Mill', but the Land Tax assessments for 1826 list 'Rutter & Co. – Mustard Mills, John Orchard – Flour ditto', indicating that the windmill was in use as a flour mill by that time even if it may not have been so previously. Another point about this entry is that the tenant is given as John Orchard, although the Rate Books continue to give the name as Joseph. Perhaps two members of the same family were running the mill in partnership, assuming that the divergence in names is not due to an error on the part of the tax or rate collectors. In 1828 the Land Tax assessments list 'Ponton' as being the owner of the entire premises, Rutter & Co. and Orchard being listed as tenants, but the following year the entry is changed to 'late Ponton', and no further mention is made of an owner.

After January 1831 there is a gap in the available records and the next mention of the mill is in the Rate Book for October 1839, where Henry Parrett, whose name also appears in the Register of Electors for 1839 and 1840, is assessed for a flour mill at a rateable value of £40, and again in the following year, this time on a value of £60. Again there is a gap in the Rate Books, but the fate of the mill is evident from the maps of the next decade. It is shown on several of them, but the area can be seen to be increasingly packed with houses, and on Greenwood's map of 1851 the route of the railway extension from Nine Elms to Waterloo is shown as passing right over the mill symbolwhich has been retained, together with the words 'Mustard Manufactory', from an earlier edition of the map.[58] Rate Book entries for 1849 and 1850 list 'Thos. Chandler – Flour Mills', but it is evident from the Post Office London Directory[59] that these refer to steam mills rather than a windmill, and it is clear from the evidence of the maps that the mill could not possibly have survived the construction of the railway extension completed in 1848, and may well have been demolished some years previously, the last map on which it is shown being that by Gardner, dated 1845.[60]

19. A watercolour entitled 'Lambeth Mill'.

Very few illustrations of this mill survive. Apart from the very badly executed illustration by Nugent mentioned in the last chapter (*fig.18*), there is an undated watercolour entitled 'Lambeth Mill' (*fig.19*),[61] which also exists as a print, showing it as a smock-mill with a stage, and also a view of St Mary's Chapel of Ease, dated 1820, in which its distant silhouette reveals a fantail attached to the rear of the cap.[62] Leigh's Panorama of the Thames of *c.*1830 shows only the sails showing above the roofs of the intervening houses and nothing of the mill itself can be seen, but by far the most authoritative illustration is a sepia drawing by John Buckler, dated 1828 and entitled 'Orchard's Flour Mill, Lambeth' (in pencil: 'Lambeth Butts'), which shows the mill in some detail (*fig.20*). The base section below the stage is apparently brick or stone-built, and contains three storeys. It is entered by means of a door at ground level, and there are adjacent outbuildings. Above the stage the body is octagonal, like the base, but with an increased batter, and this top section is covered with horizontal weatherboarding in the usual manner of a smock-mill. There are three storeys in the upper part, making six in all, and a door gives access to the stage to enable adjustments to be made to the sails. Little detail is shown of the sails themselves, but the depiction may be intended to represent a fully spread set of common sails.[63]

During the lifetime of the mill, a public house called 'The Windmill' stood on the west side of Lambeth High Street (formerly Back Lane), and a rebuilt successor to this hostelry still occupies the site, a continuing reminder of the tall windmill which once stood opposite.

20. 'Orchard's Flour Mill, Lambeth': a sepia drawing by John Buckler, dated 1828.

Site 5: Ethelred Street (formerly Gray's Walk)

The third windmill in the vicinity of Lambeth Walk was situated on a lane known in the eighteenth century as Gray's Walk, approximately corresponding to the present-day Ethelred Street which has now been truncated to make way for a new housing estate.[64]

Once again, the first map to indicate the mill is that which Rocque surveyed in 1741–5,[65] and it appears on several others during the remainder of the century, including those of Rocque dated 1769 and 1773,[66] Andrews and Drury dated 1777,[67] and Bowles dated 1786 and 1791,[68] after which there is no further cartographical record of it. As the maps published by Bowles were largely derivative, the last certain appearance of the mill is on Faden's map of 1785 where it is shown by means of the usual windmill symbol,[69] but a manuscript map also by Faden, based on an original survey of 1770, shows the mill in plan as a hatched circle situated between two buildings.[70] A map in the possession of the Duchy of Cornwall Office, purporting to be a copy of Rocque's map of 1761, shows the group of buildings in a similar way, but inspection of the original map[71] reveals no sign of any building on the site, and therefore the copy must have been made from some other unidentifiable source.

No specific entries relating to the mill are to be found in either the Lambeth Rate Books or the Land Tax assessments, indicating that the charges payable were included in amounts due from persons owning several pieces of property, the individual items not being described separately.

Pictorial records provide the only other source of information about the mill, and these are both informative and contradictory. A fine watercolour by Paul Sandby shows it standing amid a group of buildings which appear to consist of a warehouse or barn with a loading bay, some dwelling houses and various minor outbuildings. (*fig.21*). The mill itself is shown as a small brick-built tower-mill of only three storeys, with a curiously shaped cap. A stage runs round the tower at first-floor level, and a man on the stage can be seen to be operating a rope or chain hanging down from a gable at the rear of the cap in order to wind the mill. The sails are shown in some detail, and in addition to the strengthening struts behind each sail, bracing ropes can be seen joining the ends of the sails together and also to the unusually elongated windshaft projecting in front of the cap. The top of the cap consists of a box-like structure with a straight-pitched roof, and although some authorities have cast doubt on the capacity of this cap to contain a brake wheel of sufficient size, the

21. A watercolour by Paul Sandby of about 1780, showing the Lambeth 'Drug' mill.

22. An engraving by Paul Sandby, based on the previous view.

23. A watercolour entitled 'The Mill belonging to the Apothecaries Company of Lambeth'.

24. An ink and watercolour painting showing the 'Drug' mill from the opposite side. The mill at site 3 can be seen in the background.

25. A nineteenth-century engraving, clearly derived from Sandby's original watercolour.

dimensions of the cap would seem to be in keeping with the diminutive size of the tower, and as the outbuildings and associated objects seem to be of normal proportions it is reasonable to assume that the artist has depicted the mill with the same accuracy.[72] This view also exists in the form of a print, in which the entire scene is reversed, as in a mirror image, an effect which would be produced if an engraving were made directly from the original painting, without a reversal process (*fig.*22). There are two versions of this print, both signed 'P. Sandby', of which the earlier is dated 1780, thus putting the date of the original at some time previous to this. In the background of the view may be seen a wide river with a bridge and church tower, and the rural appearance of the far bank tends to suggest that this background was perhaps added to give depth to the scene rather than to adhere to strict topographical accuracy.[73] Another watercolour, apparently a sketch or uncompleted painting, shows the mill and outbuildings in a similar way but in much less detail, and here the river background is replaced by level ground (*fig.*23). In this view the artist has placed the cap at second-floor level, thus giving the mill a very squat appearance, and the stage is not shown at all.[74]

A third watercolour shows the group of buildings from the opposite side, and confirms the layout of the site and details of the mill (*fig.24*). In the background can be seen the tall windmill in Mill Street (with a rope hanging down from a winding gable at the rear of the cap, before the fantail was added) and possibly the tower of Lambeth Parish Church on the horizon. Beneath the stage there are a couple of millstones leaning against the mill beside a pile of barrels and two gear-wheels rest on the struts of the stage. Again the miller appears, this time apparently adjusting the sails by means of a rope.[75]

In Walford's *Old and New London*[B30] an engraving is printed which is claimed to be a depiction of the mill, but it is clearly a debased version of the original Sandby watercolour, and although the surroundings are shown in a credible manner, details of the mill are sadly distorted (*fig.25*). In particular, the cap seems to have shrunk to include only the oblong structure, sitting awkwardly on the rounded top of the mill, with the winding gable projecting from the mill body instead of from the cap! This picture has subsequently been published in books and newspaper articles describing the mill, but little veracity can be claimed for it.[76]

The Lambeth 'Drug' mill

In describing the mill illustrated in his book, Walford makes the remarkable claim that it had belonged to the Apothecaries Company and was used by them 'for grinding and pounding their drugs, &c.', and although he gives no source for his information this statement has been repeated by all later writers on the mill, accepting it at its face value. R. Thurston Hopkins, for instance,[B13] repeats the statement and describes the construction as being of 'wattled walls of plaster fully three feet thick', although Walford on his part has it 'built chiefly of wood', but both of these descriptions may be dismissed as sheer speculation engendered by the inaccuracy of the engraving. Sandby's original watercolour shows quite clearly that the tower was brick-built and rather small in its over-all dimensions, thus rendering three-foot thick walls to be somewhat excessive (*see fig.21*). Hopkins also considers the lantern projecting from the top of the warehouse building to be one of the Lambeth shot towers, but in addition to the fact that the original views show that it was not a free-standing tower, it bears no resemblance whatever to either of the two shot towers, and in any case, the first of these was not built until *c.*1789, several years after the date of publication of Sandby's print based on his earlier watercolour of the scene.

In his *London Memories*,[B26] Heckethorn repeats Walford's remarks almost verbatim, and several later writers, not realizing that there was more than one windmill near Lambeth Walk, have located the so-called 'Drug' mill in Mill Street (see Site 3, described above). Coming further up to date, Farries and Mason have recorded Walford's statement,[B19] together with a note from the L.C.C. Survey of London to the effect that there is no evidence to indicate ownership by the Apothecaries Company, but then go on to accept any illustration inscribed 'Apothecaries Drug Mill' as being a picture of the windmill in Mill Street, even though several different types of mill are clearly depicted. Practically every illustration of a Lambeth windmill in public collections is labelled 'Drug Mill', which is probably due to the activities of over-zealous archivists and librarians who have read the books mentioned above and have accepted the designation without further enquiry and applied it indiscriminately to the illustrations in hand.

Although Walford's statement seems to be too extraordinary to have been invented without some basis in fact, careful inspection of the records of the Society of Apothecaries[77] has failed to reveal any evidence for a connexion between the Society and a windmill in Lambeth. The existing records consist of court books, account books, and inventories of property, which deal in some detail with the administration of the business of the Society and the maintenance of its various properties, but there is no mention of any item in the parish of Lambeth, let alone a windmill. According to these records, the Society possessed its own facilities for grinding the substances required by its members, adjacent to the Apothecaries Hall in Blackfriars, and it is possible that if large quantities had to be dealt with, part of the work could have been passed out to external contractors to be ground in various types of mill.

Mustard is listed as one of the products, and it will be remembered that the windmill near Broad Street (Site 4) was connected with a factory described as 'Mustard manufactory', which may well have taken work as sub-contractor to the Apothecaries Society, thus giving rise to the story of a 'Drug' mill near Lambeth Walk. An alternative explanation is that the inscription 'The Mill belonging to the Apothecaries Company of Lambeth' on one of the views (*fig.23*) suggests that this was perhaps a local organization not connected with the Society of Apothecaries proper, but as the latter was incorporated by Royal Charter it is hardly likely that two such organizations could have existed simultaneously, and the Lambeth concern may therefore have been a commercial company engaged in the production of medicines and powders.

Because of this indiscriminate description of views of Lambeth windmills as being of the 'Apothecaries Drug Mill', some confusion has arisen as to the relative location of the three mills depicted in this area. Sandby's original watercolour (*fig.21*) carries the inscription 'Near Cuper's Bridge' and in another hand 'No! – Lambeth Walk', but from the evidence of the view shown in fig. 24, where the tall smock-mill appears in the background, the Cuper's Bridge location (i.e. close to the present Waterloo Bridge) must be incorrect. Moreover, the latter view is entitled 'Camberwell Mill', thus adding even further to the confusion!

Although later writers have tended to locate the mill depicted by Sandby in Mill Street, it is quite clear that Site 3 was occupied by the large mill shown in fig.15. The view by H. Pyall shows the surroundings with some accuracy and fixes the site of the mill in relation to Lambeth Church tower, which appears in the background, and the view by John Buckler is inscribed 'Wind Mill in Mill Street, Lambeth' (*fig.17*). Buckler was a trained architect and made a number of drawings of buildings in this area, many of which are now in public collections, which are all correctly described with topographical accuracy, so we may accept his caption to the drawing as confirmation of the location indicated by Pyall's painting. Farries and Mason, however, place the small 'Drug' mill at the Mill Street site, and account for the variation in the type of mill depicted in the various illustrations by assuming that it must have been rebuilt at some time, but there is no reason to suppose that any rebuilding took place at this site, as the tall mill appears in several views of Lambeth Palace, commencing in 1735, whereas the much smaller 'Drug' mill would hardly have been visible above the surrounding trees.

Having allocated the Mill Street site to the 'Drug' mill, Farries and Mason locate the large smock-mill at Site 4, near Broad Street, but again Buckler's drawing of 'Orchard's Flour Mill, Lambeth Butts' (which was not available to these authors) shows a smaller octagonal windmill of quite different design (*fig.20*) and the consistent listing of Joseph Orchard in the Rate Books and Land Tax assessments confirms the representation of the mill occupying this site. The Rate Books assess the value of the Mill Street site as being much higher than that of the Broad Street mill, and Leigh's Panorama of the Thames of *c.*1830 shows the tall mill at Site 3 towering behind Lambeth Church, while only the sails of the Broad Street mill show above the rooftops, both of which factors would hardly be the case if the positions of the two mills were reversed.

Site 5, therefore, provides the only remaining location for the 'Drug' mill depicted by Sandby, and in the absence of other evidence it must be assumed that this was where it stood, and that the river background shown in fig.21 is ficticious. Comparing figs.21 and 24, it is clear that the warehouse building could not both face the Thames and also be aligned towards Mill Street in the manner shown. One of them must be wrong, and as the site of the large smock-mill has been confirmed from other sources, the river background in fig.21 must have been inserted to complete the view, possibly being added later when the watercolour painting was done over an original pencil sketch. The simplified representation of the plan of the site on the two maps mentioned in the last chapter does not correspond in detail with the rather complex juxtaposition of buildings shown in the illustrations, but it is sufficient to indicate that the windmill stood amongst a group of outbuildings of some size and therefore may well have been the so-called Lambeth 'Drug' mill.

Footnote, August 1971: While this book was in the press, new evidence came to light which both confirms the site of the mill and provides an explanation of its connexion with the Society of Apothecaries. In the course of research into a quite different topic, Miss Rhoda Edwards, Lambeth's Assistant Archivist, discovered a reference to the mill in the Court Rolls of the Manor of Lambeth, which are filed on microfilm in the Surrey Collection at Minet Public Library, and from this it has been possible to trace other entries relating to the mill. At a court held on 10 July 1759, the Reverend John Dry of Thames Ditton, Surrey, surrendered to the Lord of the Manor

> All that Customary Messuage or Tenement Windmill and piece or parcel of Ground containing by Estimation Two Roods and Fourteen perches holden at the yearly Rent of 6d in the Tenure or Occupation of George Rutt of Friday Street London Druggist And also all that customary piece or parcel of Garden Ground adjoining thereto containing by Estimation One Acre Two Roods and Twenty two Perches abutting North, on Three Coney Walk, South upon Garden Ground in the Occupation of Edward Manning, East on Garden Ground now in the Occupation of Thomas Riddlesdale and John Riddlesdale and West on a Walk called Grays

> Walk and holden at the yearly Rent of 3^{d} lying in Cotman's Field Together with all and every the Appurtenances to the aforesaid premises belonging or in anywise appertaining

and on 8 May the following year, after some dispute as to whether John Dry was legally entitled to the property, George Rutt was formally admitted as tenant. The next reference occurs in 1778, when a court sitting on 22 May heard of the death of George Rutt, and admitted the trustees of his estate to the tenancy of the windmill and its associated property. In addition to Rutt's widow, the trustees appointed by his will included Samuel Etheridge of Lewisham, a hosier named John Howell of Newgate Street, London, and John Field, an apothecary of the same address. On 9 May 1791 the death of John Field was reported to the Court and although arrangements were made to deal with the administration of his freehold properties, no mention was made of his obligations as one of the trustees of Rutt's estate, and there are apparently no further references to the windmill in the Court Rolls. It is clear however that the mill was in the tenancy of persons following the trades of druggist and apothecary for a considerable number of years and was doubtless used by them for the preparation of their materials. It seems that in its earlier years the mill may well have been used for grinding materials in the manufacture of pottery, as the following advertisement which appeared in *The Country Journal, or, The Craftsman* for 10 January 1730 presumably refers to this site (again, thanks are due to Miss Edwards for drawing this to my attention):

> To be Lett at Lambeth in Surrey a Potter's Wind-Mill; together with a small House and Garden thereunto belonging: And also Two Acres and an half of Garden-Ground. Enquire at Lambeth Wells, or at Mr Dry's Lodgings, at Mr Bray's House in Pearl-Court in White Fryars.

Site 6: Nine Elms (Randall's mill)

Windmills have existed at Nine Elms since early times, the Thames-side location being ideal for catching the wind without obstruction from neighbouring buildings. The district itself is divided between the two parishes of Battersea and Lambeth, and although the mill to be discussed lay on the Battersea side of the border, it is included in this survey as being of interest to Lambeth readers.

The Battersea Rate Books have many entries for mills, and some difficulty is experienced in interpreting them as no precise locations or descriptions are given. In the account books kept by the Battersea Overseers of the Poor there is mention of a 'Corne Windmill' and a 'Coloure Windmill' in 1684 and subsequent years, and although they are listed under the heading 'Nine Elms', their precise location is uncertain.[78] A survey of the Manor and Parish of Battersea dated 1763 shows a windmill marked as 'Colour Mill' in Battersea Fields, a good distance from Nine Elms proper, which may well be the one referred to in the account books.[79] Writing in *Our Lady of Batersey*,[B31] J. G. Taylor claims a third windmill 'for grinding white lead' in addition to those mentioned above, and that all three are described as 'the first mills near London' in the Overseers' Accounts for 1675, but inspection of the document concerned has failed to produce any verification of this. A pamphlet entitled *Historic Battersea*[B22] puts the date of mention of the mills in the Accounts at 1649, but again this cannot be verified. The account book for 1688 has the entry 'Mr. De Wilder at ye colour mill', and in 1751 there is 'Addison – Colour Mill'.

The first direct evidence of a windmill existing on the present site comes from Rocque's map of 1762, which employs the conventional post-mill symbol to indicate the mill,[80] and subsequent maps confirm the site as being on the Thames bank, right at the water's edge, on the western side of the mouth of a creek.[81] This creek is in fact the outlet of the River Effra, which rises near Dulwich and has served as a boundary between the parishes of Lambeth and Battersea since Saxon times.

Commencing in 1780, Land Tax assessments show that the mill was in the ownership of William Saunders and was leased to Randall & Co., whose name became permanently associated with the mill, even after their lease had expired.[82] A watercolour by John Cleveley (1747–86) entitled 'View of the Thames near Vauxhall' shows it as a stone-built tower-mill standing on a wharf jutting out onto the Thames foreshore(*fig.26*). It apparently has three floors above the wharf level (each with oblong windows), and there

26. *'View of the Thames near Vauxhall': a watercolour by John Cleveley, showing Randall's mill at Nine Elms.*

27. *Detail from an eighteenth-century 'View of the River from Vauxhall Stairs' by W. Marlow, showing Randall's mill and associated buildings.*

is a semi-spherical cap surmounted by a finial. The four anti-clockwise sails are of the common variety, and are shown fully spread on this view.[83] An eighteenth-century plan of Battersea parish shows the windmill marked as 'Mr. Ponton's Corn Mill',[84] although the Land Tax assessments continue to give the proprietor as 'Saunders', but in 1798 Thomas Ponton appears as proprietor in the assessments, so perhaps previously Saunders was the lessee and Randall & Co. were sub-tenants to Saunders. The taxable value of the mill is given as £56.

The first mention of the mill in the Rate Books occurs in 1792, and the details given in the Land Tax assessments are confirmed, although there are many previous entries, from 1751 onwards, for a John Randall at the 'Docks' section of Nine Elms, without mention of a windmill.[85] An aquatint by C. Tomkins entitled 'Millbank looking Eastward' published in June 1796 shows the mill with four *clockwise* common sails, and a door can be seen in the mill body facing the river, giving access to the wharf. Another door is apparent about halfway down the south side of the wharf itself, which appears to be weather-boarded rather than stone-built, suggesting that the portion of the wharf below ground level was used for storage and loading purposes.[86] This timber construction of the wharf is confirmed by a painting by Daniel Turner dated 1801, which also shows vertical rows of porthole-like windows in the mill body and a ball replacing the finial on top of the cap. A gable projecting from the cap affords housing for the winding gear, but no wheels or chains can be seen in this illustration.[87]

In 1799 'Saunders' replaces Ponton as proprietor in the Land Tax books, but in 1801 'Thos Ponton do. Saunders' appears. In 1802 the entry reads 'John Saunders', possibly a son or other relative of the original William Saunders, and Randalls remain the occupiers until 1808, when we read 'Mill Empty', with the corresponding assessment of £74 crossed out, but in the Rate Books Randall & Co. disappear as early as 1805, where the occupier of the mill is listed as 'Shee, late Randall'. From this date onwards a fair amount of confusion reigns in both sets of accounts, due to the failure of both to specify the exact location and nature of the mills referred to: in addition to at least three windmills in the Nine Elms area there was also a water-mill, and all are referred to simply as 'Mill'. To add further to the confusion, some books do not describe the property assessed at all, consisting merely of lists of names, and it is only at a much later date when a rate book with provision for more detailed entries came into use that further information is available, so for an account of the history of the mill in these intervening years we must turn to pictorial records.

28. Randall's mill as depicted in the 1808 Perambulation of the Parish of Lambeth.

Many views of the river dating from the early nineteenth century show the mill, some as the main feature, and others as part of the scenic background. A view from Millbank by T. Baxter dated 1802 confirms the details shown in the illustrations mentioned previously, and adds square windows in the lower part below the wharf, the double doors being shown wide open. The door to the upper part can be seen to have a small triangular gable above it.[88] In 1808 the mill was marked on a manuscript map accompanying a Perambulation of the Parish of Lambeth, where it is drawn as a smock-mill (*fig.28*), but as the depiction is very similar to that used for the Brixton Hill windmill, it was probably not meant to be an exact representation, but merely as an indication of the presence of the mill as a landmark.[89] An idea of the rather squat appearance of the tower can be gained from an engraving by W. Cooke from a drawing by S. Owen, published in May 1809, and this picture also indicates the amount of 'weathering' employed on the sails (*fig.29*). The method of fixing the sails by directly mortising the stocks through the wooden windshaft, instead of using the later cast-iron poll-end or cross, can be clearly seen.[90] Most views show the mill as seen from the river, but an etching by Samuel Prout dated October 1815 shows the landward side, with a woodyard in the foreground (*fig.30*). The mill

29. An engraving of Randall's mill, from a drawing by S. Owen, first published in May 1809.

does not seem to be in active use, judging by the appearance of the sails, but a good view can be obtained of the winding gear, consisting of a large spoked wheel tangential to the mill body, protected from the elements by a gable projecting from the cap, with an endless chain hanging down to ground level.[91] A view of Vauxhall Bridge, dated 1816, shows Randall's mill in the background; three of the four sails can be seen – two adjacent ones of the clockwise variety, and the third anti-clockwise![92]

Of the many illustrations depicting the mill at this period in the form described above, the last is a sketch dated August 1820.[93] After this date some changes appear. A lithograph by F. E. Martens, dated October 1829,[94] shows the mill (although rather inaccurately drawn) with the addition of a fantail and double-shuttered sails (*fig.31*), and the fantail can also be seen in Leigh's 'Panorama of the Thames' of *c.*1830 (*fig.32*).[95] The exact nature of the transformation can be seen in a fine watercolour entitled 'Mill near Vauxhall – study from nature' by J. Varley, dated July 1830 (*fig.33*). It clearly shows the new double-shuttered sails and a high eight-bladed fantail supported by an outrigger structure which has been built onto the back of the previous winding gear gable. A long pole strengthened by struts hangs down to the wharf level to enable the cap to be turned by hand without the aid of the fantail (as at the Berney Arms mill, Norfolk), but the winding wheel and chain remain as before. Perhaps this wheel was now disengaged from the rack, rendering the tailpole necessary in order to move the cap to a new direction in the event of a lull in the wind or in case of a failure of the fantail mechanism.[96] Another view of the mill by the same artist shows the front of the sails (*fig.34*), and the absence of a striking rod and spider mechanism indicates that the pre-set spring shutter type of sail was in use.[97] Both these views show four floors to the mill (in terms of external windows), and the tower is painted a dark colour with a light horizontal band at third-floor level.

32. Detail from Leigh's 'Panorama of the Thames' of c. 1830, showing the high fantail fitted to Randall's mill.

30. 'Near Vauxhall': an etching by Samuel Prout, dated October 1815.
31. 'Near Vauxhall Bridge': a lithograph by F. E. Martens, published in October 1829.

33. 'Mill near Vauxhall—study from nature': a watercolour by J. Varley, dated July 1830.

34. Another view of Randall's mill by J. Varley, showing the spring shutter sails.

Two illustrations by Thomas Hollis, now in the Guildhall Library (and labelled, predictably, 'Apothecaries Drug Mill'), confirm the details shown in the Varley paintings, and also show a small gable projecting from the front of the cap to shield the windshaft, similar to the larger one at the rear which has already been mentioned. One illustration is dated 1833 and is marked on the reverse 'Mill at Nine Elms, Vauxhall, as it was formerly', probably meaning that the drawing was done some time prior to that date (or else this remark has been added by a later hand), and the other, a watercolour, shows the mill body as being painted white, and the cap black (*fig. 35*).

Returning to the Rate Books, the next date at which an entry definitely associated with this mill appears is 1835, when Charles Knott is listed as the occupier, the owner still being given as 'Saunders'. Similar entries for the following years occur, and the book for 1838, besides giving the rateable value (£50), also records the extent of the property, which was 28 perches. Here another gap in the Rate Books occurs, and the next series of entries appears in the years 1839–42, where the owner remains the same, but the tenancy has passed to Messrs Francis & Co. The last entry of all which refers to the mill appears in the Rate Book for 1843, where, although Francis & Co. still appear as tenants, all the figures relating to the entry have been erased and the proprietor is now given as 'South Western Railway Company'. Thus yet another windmill became a victim of the mid-nineteenth century railway expansion, probably in the course of changes in the railway layout at Nine Elms terminus, following the extension of the line to Waterloo. At the time of writing, the surrounding area is again under development, this time to accommodate the premises of the new Covent Garden Market. This project involves the demolition of buildings and destruction of equipment belonging to the railway system, but any vestiges of the windmill have long since disappeared.

The last map to show the mill is that of Gardner, published in 1845,[98] and Cruchley's map of 1851 merely marks the 'Railway and Windmill Piers', the mill itself no longer being recorded.[99] An earlier manuscript map of 1836 relating to a purchase of land by the South Western Railway Company shows a railway line running from Nine Elms station, across both Nine Elms Lane and the lane leading to the windmill, to the edge of a wharf, which had apparently been built out into the mouth of the creek to the east of the mill, and the demise of the mill was probably due to the extension of these wharf facilities by the railway company.[100] As a matter of further interest, this map shows a public house called the Windmill standing on Nine Elms Lane by the lane leading to the mill itself, and the associated Book of Reference gives John Saunders as owner, Charles Francis as lessee, and James Purdy as occupier of the building.

It is not clear whether Randall's mill was used for milling corn, or for some other purpose. In his *London Memories*,[B26] Heckethorn states that it was for 'grinding corn', and Johnstone's *London Commercial*

35. A watercolour by Thomas Hollis, inscribed 'Apothecaries mill, Nine Elms'.

Guide and Street Directory of August 1817 lists John Cornwall, miller and corn dealer, at Nine Elms, but whether he was operating at Randall's mill is not known.[101] On the other hand, several maps of the early nineteenth century mark the immediately surrounding area as 'Cement factory', 'Cement wharf', or 'Roman Cement Manufactory', and it is possible that the windmill was in some way connected with this industry, at least for part of its life. Among the papers of the engineer John Smeaton are some drawings referring to a China Mill at Nine Elms, Vauxhall,[102] but again, whether this was Randall's mill is not certain. There are five drawings in all, the first two being plans for the tub machinery in the mill; as first designed, and then as finally installed. The first shows an arrangement of eight overlapping circles with a central wheel (apparently a trundle cog wheel) supported by a braced octagonal frame (*fig. 36*) and the second is similar but with a square instead of octagonal frame, although exactly how the mechanism was employed for grinding is not clear (*fig. 37*). The third drawing is a scale of weathering for the sails, while the fourth is a front view of the four sails (*fig. 38*). They are common sails, anti-clockwise, with a triangular portion on the leading side of the whip, but no scale of size is given. The last drawing shows 'A Projection for Weathering Windmill Sails in the manner of those at Vauxhall', and gives a scale drawing of such a sail, which is shown to be 30 ft long (sail only, excluding whip) on the trailing side, while the triangular portion on the leading edge is 5 ft wide at the tip of the sail (*fig. 39*). None of these drawings is dated, but there is a note with them which states that 'In the list of mills executed by Mr. Smeaton, Nine Elms China Mill wind, is the first of all', and they can therefore be assigned to about the middle of the eighteenth century.

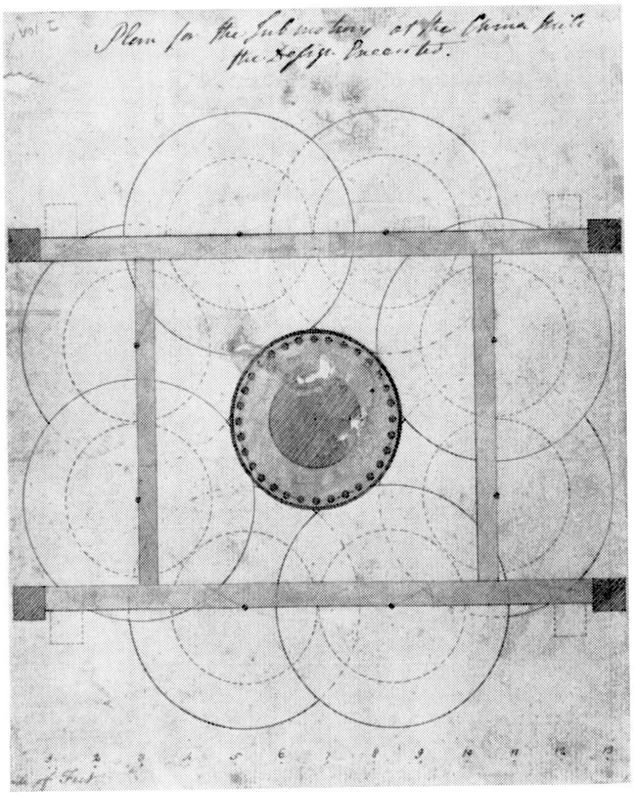

37. 'Plan for the Tub machinery at the China Mill—the Design Executed': a subsequent drawing by Smeaton.

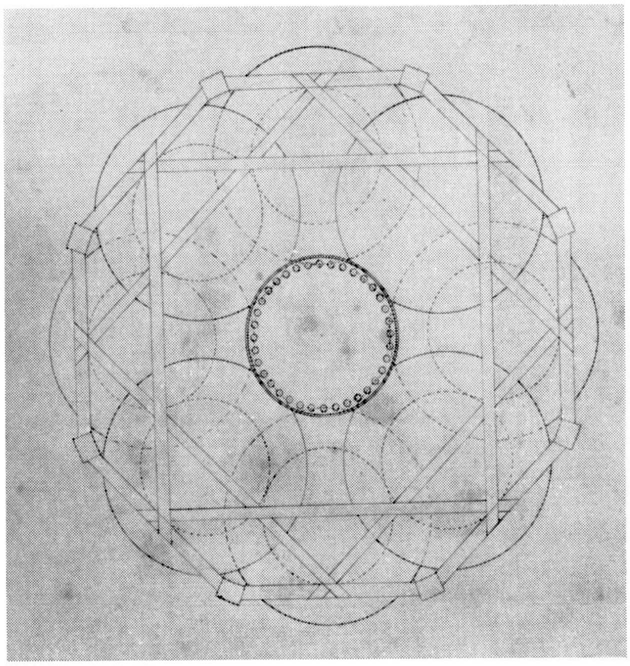

36. 'Plan for the Tubbs at Vauxhall China mill': a drawing amongst the papers of John Smeaton.

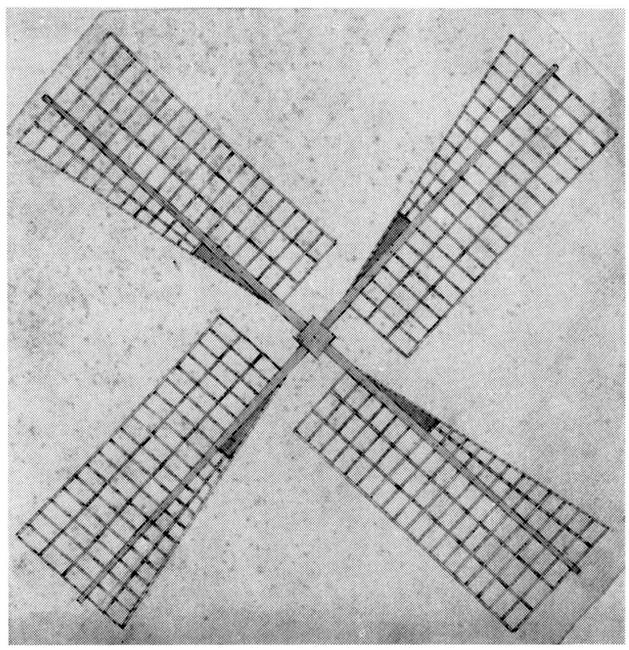

38. Design for a set of common sails by John Smeaton.

39. 'A Projection for Weathering Windmill Sails in the manner of those at Vauxhall' by John Smeaton.

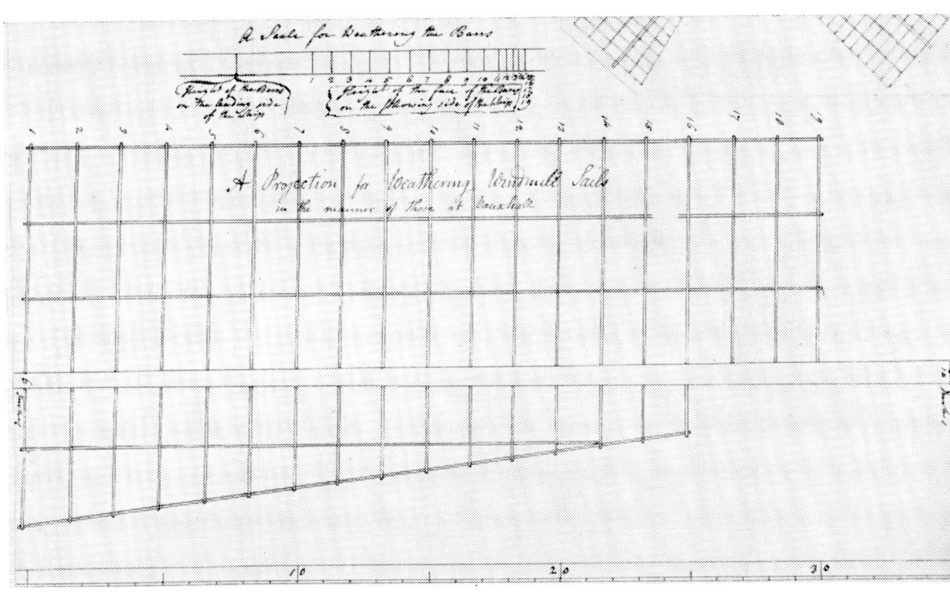

Site 7: Clapham Common

Although there are no maps or illustrations existing today which show a windmill at Clapham, it is most fortunate that several remarkable documents have survived the vicissitudes of time, and enable a picture of what the mill was like to be built up.

Most important of these documents are several leases which relate to the windmill itself; the earliest legal documents now existing which refer to a specific windmill in the Lambeth area. All are written on parchment, and give the names and occupations of both parties, and in addition there are schedules which list the contents of the mill.[103] The first is dated 14 July 1631, and is between Henry Atkins, of London, 'Doctor in Physick, one of His Majestie's Physicians in Ordinary', and Richard Jenman, of Clapham, yeoman, and relates to 'all that Windmill, dwelling house, and little plott of ground about the said house', lately in the occupation of one Christopher Godfrey, and now occupied by Jenman. The term of the lease is twenty-one years, and the rental is given as £8 per annum. Attached to the lease is a schedule of great interest, which lists the contents of the mill and gives information regarding its construction, and although the document has not survived the ages very well and is partly illegible, its content is apparently the same as a much better preserved schedule attached to a later lease (*fig.40*). It reads:

THE SCHEDULE of the goods things implements and utensills in and about the said Mill demised by the Indenture annexed vizt:

In the Upper roome of the Mill

Imprimis One paire of Millstones of ffive foote broad the upper stone being six inches thick att ye Skirts.
Ite. The Tunne about the stones with a cover over it
Ite. The Binn shoe and ladder to the stones
Ite. The Twindle* head with an iron spindle & staves
Ite. The Cogg wheel with new Coggs and the Breake
Ite. The winch and wheel to winde Sacks upp to the Mill
Ite. The Collar Brasse of the best bell mettall. It cost 12p. per li and is in weight Twenty ffower pounds.
Ite. The Lighter
Ite. The Axtree that carryeth the Sayles with Iron ribbs att the one end and a Brasse gudgeon att the other waighing 12lb and it cost 12p. per lb with an Iron Gudgeon and Hoope.
Ite. A Paire of Stayres to the said roome
Ite. The roome well florred with oke boards

In the Lower roome

Imprimis The Trough and the Spoute
Ite. one wooden leaver
Ite. one under spindle of Iron running upon a steele Gudgeon
Ite. Three Bushells and one Peck waight
Ite. The Roome well floored with oake boards
Ite. a lock and Key to the Door.

* *i.e. Trundle*

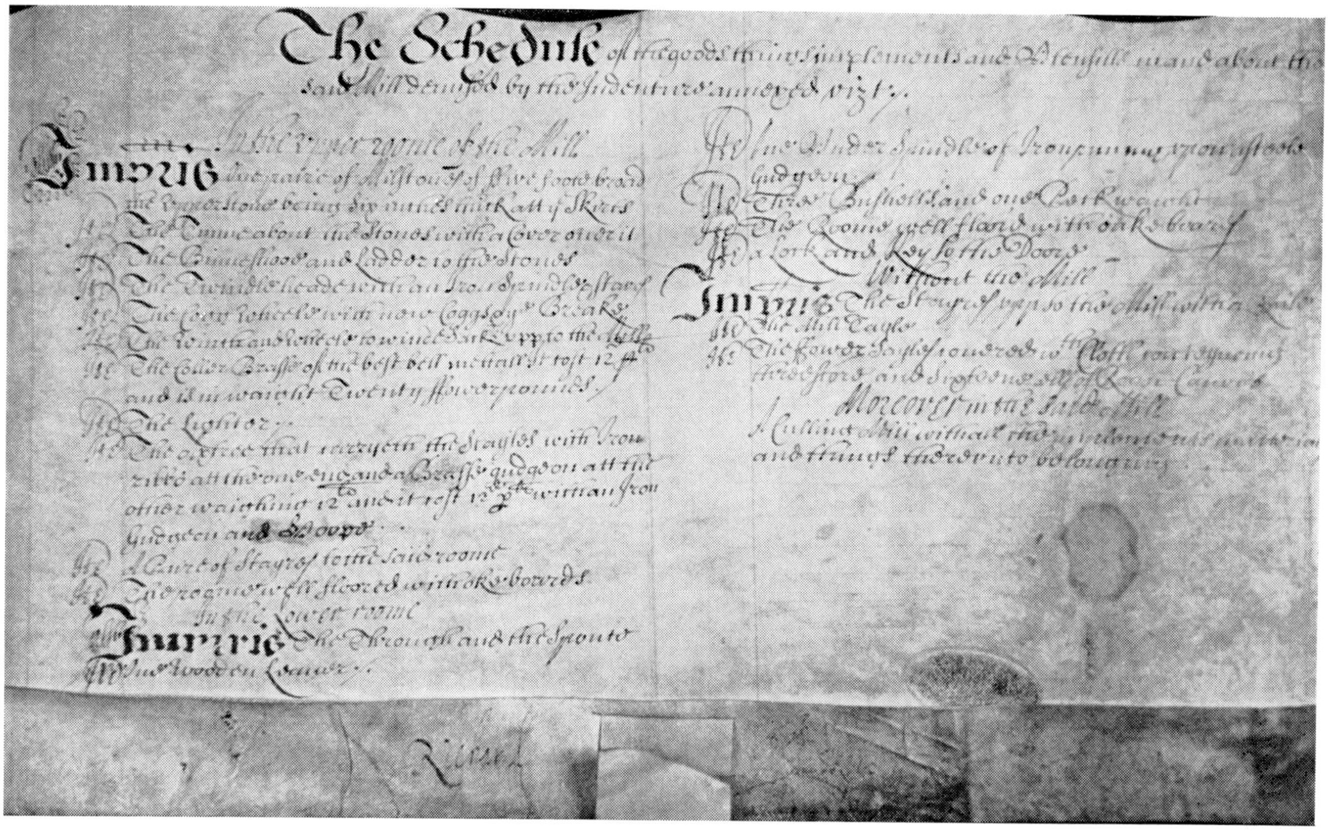

40. Schedule attached to a lease of 1651 relating to Clapham windmill, listing the 'goods things implements and utensils in and about the said Mill'.

Without the Mill
Imprimis the Stayres upp to the Mill with a raile
Ite. The Mill Tayle
Ite. The ffower Sayles covered with Cloth conteyning three score and sixteene ells of Roan Canvas.

Towards the end of the term of this tenancy, a new lease was taken out, this time between Richard Atkins, of Much Hadham, Herts, and Andrew Weston, of Clapham, on 10 June 1651. The rental and period of lease are the same as before, but there is a slight modification in the description of the property, which now runs 'All that Windmill, dwelling house and little plot of ground about the said house now severed from the common'. This is a useful piece of information, indicating that the mill stood on the Common, and that presumably a fence of some kind had been erected round the plot of ground in the time since the first lease was granted. This new lease ran for only seven years, and on 7 March 1658 a further lease was drawn up between Richard Atkins and Thos. Chrenshawe of Clapham. The term of this lease was to be sixteen years, and the rent had been increased to £10 per annum. Attached to the lease is a schedule which gives the contents of the mill as listed above, with the addition of the following paragraph:

Moreover in the said Mill
A Culling Mill with all the implements materials and things thereunto belonging.

This indicates that an extra pair of stones, of the Cologne, or Culling type, had been installed in the mill, which probably accounts for the increased rental. However, this lease also did not run for its full term, for the next one is dated 12 January 1666 and is between Atkins and John Letten of St Mary Whitechapel, for a term of twenty-one years. The property is briefly described as 'mill with its appurtenances in the parish of Clapham', and it seems that there must have been some deterioration of the condition of the mill during the previous tenancy, as the rent is now only a peppercorn for the first two years, and £3 per annum for the remainder. In addition, the attached schedule is quite different from the preceding ones, and mentions only the equipment associated with the Cologne stones; the main stones presumably having fallen into disuse. The schedule reads:

One paire of Cullingstones a foot thick apeece and six foot & nyne inches over.
One paire of old stones att the head of ye mill.
One paire of scales with a wooden beame & Three half hundredweights one with a ring and two without and one 14 pound weight.
One cable, one sledge, and one crow.
One collar brass, two foot brasses, and one stepp brass at the little end of the axell tree and two mill bills.

The 'sledge' and 'crow' referred to are of course a sledgehammer and crowbar, but the list is chiefly remarkable for the very large size given for the pair of Cologne stones. The mention of a pair of old stones at the head of the mill lends weight to the supposition that the original stones and associated mechanism had become disused by this time. No subsequent leases relating to the windmill exist today, but an undated (but post-1660) lists of rents in the Manor of Clapham gives the mill house at a rental of £8 p.a., without mention of the mill itself.[104]

Further information about the mill comes from a set of papers prepared during one of the recurring boundary disputes between Clapham and Battersea parishes, both of which claimed rights on Clapham

Common.[105] This particular action was brought in the year 1718, and various inhabitants of both parishes put forward their views on the precise location of the boundary. One of the witnesses declared that 'Many years ago there was horseracing by ye Gentlemen of Clapham both on ye East and Westward parts of ye said common & to such end about 5 posts were set up at certain distances beginning at ye Windmill on ye East part of ye said common and from thence . . . round the same to ye windmill aforesaid'. The father of another witness remembered 'a windmill then formerly standing on the hill in Ballam Wood Lane, and that the mill was removed thence to the east part of the common to the Windmill Hill where the starting post for the races was'.

It thus appears from these documents that the windmill stood on the 'east' side of the Common (i.e. what is now called South Side) at some time during the seventeenth century, and in all probability its site coincided with that of the present Windmill Inn, which dates at least from the early eighteenth century.[106] The author of a pamphlet entitled *Clapham: old yet ever new*[B21] published in 1937, claimed that 'An old map (dated 1680) shows the windmill which gave the Inn its name', but no trace of such a map can now be found. The earliest maps now existing of Clapham date from the eighteenth century and show the Windmill Inn, but not the mill itself. The abbreviation of the name of the inn to simply 'Windmill' on some of these maps may have given rise to some confusion. Farries and Mason record that in 1665 Thomas Crenshaw, miller (presumably the tenant mentioned in the lease of 1658) also kept an alehouse.[B19]

The story of the removal of the mill from an original site in Balham Wood Lane (now Nightingale Lane) is interesting. Such removals were not uncommon as there are several recorded instances of post-mills being moved several miles to new sites, but the fact that the mill was said to have been moved to a place on the common known as the Windmill Hill raises the possibility that a mill had previously stood on the new site.

Two documents, dated 25 November 1611 and 25 November 1618, relating to transfers of ownership of the Manor of Clapham, mention only one windmill amongst the various items of the estate,[107] but there are other references to two windmills standing simultaneously in the Manor. In *The Chronicles of Clapham*,[B20] published in 1929, J. H. Michael Burgess states that when Henry Atkins bought the Manor of Clapham in 1616, there were, and had been for a considerable time two windmills in existence, and two Feet of Fines documents dated 1617 and 1631 mentioned by Farries and Mason also record the fact of two windmills in the Manor,[B19] but none of these sources specifies the sites on which the mills stood.

Site 8: Blenheim Gardens, Brixton (Ashby's mill)

The only windmill now surviving in Lambeth stands today in a pleasant garden just off Brixton Hill, providing a welcome retreat from the roar of the motor traffic on the main road.[108] The land on which it stands once formed part of the Manor of Stockwell, and was purchased in 1802 by a Southwark merchant named Christopher Chryssell Hall at an auction of lands belonging to the manor.[B37] In 1816 the windmill was constructed and was leased in November the following year to John Ashby, a miller of Brixton Hill, together with two acres of land and some associated buildings, and the firm of Ashby & Sons operated the mill for the whole of its working life. According to an old inhabitant of Brixton, the Ashby family were Irish Quakers, and his uncle whom they had brought with them as a boy had become manager of the mill in later years.[109]

In August 1819 John Ashby took out a mortgage on the entire property, which was not redeemed until much later in the century. His name first appears in Pigot's Directory for 1823–4, where he is listed under 'Brixton: millers',[110] and the Lambeth Rate Book for 1824 lists 'John Ashby for House and Mill' at a rateable value of £30, under the heading of Rush Common, which was the name of the land bordering Brixton Hill in those days. Robson's Directory of 1838 gives the entry as 'John Ashby – Lower Mills, Brixton', to distinguish the windmill from the one which stood further up Brixton Hill, to be described in the next chapter.[111] The Census of 1841 records the Ashby family as consisting of John, aged 65, Hannah, his wife, aged 60, and three of their children; Aaron (30), Julia (25), and Joshua (20). Of these, only Joshua was born in the County of Surrey, but no information is given about the origins of the family.[112]

Some interesting information about the mill in its early years may be found in the ledgers of William Ashby, millwright, of Westerham, Kent.[113] There is no evidence that there was any connexion between the two families of the same name, especially as the Brixton Ashby's were reputed to have come from Ireland, but as they were all in the same business of milling there may well have been some relationship. Two other Ashbys are also mentioned in the ledgers (apart from William's own sons) – a John Ashby at Brasted, Kent, and yet another John Ashby, working at Carshalton in Surrey. This latter John was a son of the John Ashby at Brixton, and although charges for work carried out at Carshalton were often included in the accounts for the Brixton mill, each item of work was carefully specified, and to avoid any confusion between the names, work done at Brixton was clearly marked: 'John Ashby of Brixton'.

The first entry in the ledger dates from 10 August 1822, when William Ashby's son Thomas worked for seven days at Brixton at a cost of £1–8–0. On the first of January the following year there was a charge of £2–10–0 interest on a sum of £50, which could represent an unpaid debt, or, more likely, part of a sum of money borrowed by John Ashby to enable him to set himself up in the milling business, paying the balance back in instalments. Also on this date there was a charge of £1–16–0, being 'Whites Bill for youse of Thrashing Machine'.

On 16 July 1825 there was a charge of £3–3–4 for drawing up a copy of a lease of Erith Mill, followed by a list of several charges for grinding various quantities at Erith. The most likely explanation for John Ashby taking a lease on another mill and William Ashby being concerned in it is that perhaps some serious technical problems had arisen at Brixton, necessitating a major overhaul of the machinery during which grinding facilities would have had to be found elsewhere, but there is no charge mentioned in the ledger for any work carried out at Brixton that year, so perhaps the lease was due to a large amount of extra work being taken on which could not be coped with at Brixton, and William Ashby acted as an intermediary between the two millers concerned.

In April 1826 various items of work were carried out, including installing a new stone spindle, stone nut, brasses, quants and tenter irons and associated machinery, while in 1827 there were various labour charges in March and July, and in September some major work was carried out on the sails, involving fitting two quadrants, a cross, a rack and pinion mechanism, a 6 ft rigger with iron arms, and drilling and centring various items. This probably refers to the conversion of a pair of spring sails into patent sails by adding a striking rod system to enable the amount of wind spilled by the shutters to be adjusted without stopping the mill.

41. Ashby's mill, Brixton: an early watercolour.

The following month a new wallower wheel was fitted at a cost of £22, described as 'A new Iron Wollar 3 ft. 6 in. dim. Pitchd & Trimd with Iron face ring for hoisting sacks etc. – geared with wood'. It is interesting to note that wooden gear teeth were still being fitted in windmills at this date, even though the main part of the wheel was made of iron. In 1828 and subsequent years there were entries for various items of maintenance, and in 1834 seventy-two cogs for the brake wheel were fitted, which must again have been made of wood. At the same time a 'Ragulator Brass' was installed, together with 'An Iron fork forg'd & file'd up to fitt Ragulator Brass'.

In December 1836 John Ashby brought an action in the Court of King's Bench against the Company of Proprietors of the Lambeth Water-Works for stopping up a watercourse and thus injuring his premises, and William Ashby's Day Book records that he travelled to Brixton on 2 December to act as a witness the next day, and also on 10 December, where he notes 'Self to Guild-Hall as witness & Return'd on the 15th, being detain'd 5 days as witness'. Details of his expenses relating to lodgings and coach fares are given. The case was heard by the court on Saturday, 3 December, and judgement was given in favour of Ashby, subject to a view of the premises, which accounts for William Ashby's second journey to London to act as witness.[114]

In August 1837 a new stone nut $12\frac{3}{4}$ in. in diameter was fitted to the mill, together with 'a rigger $18\frac{1}{2}$ inches diameter, bolted on with $4 \times \frac{3}{8}$ bolts and 2 oak wedges to hang it on the spindle with', and various other items including twenty-five cogs for the stone nut at 7*d* each, 12 ft of $\frac{7}{8}$ in. Elm board at 5*d* a foot, and 13 ft of $1\frac{1}{8}$ in. board at 6*d* per foot. 'A cast Iron carriage with a bracket & key' was also supplied at a cost of £4, and the work included '4 small spindles with 2 barrens (*i.e. bearings*) to each of them turned to $\frac{1}{2}$ size'. There are also entries listing work carried out in the years 1838–40, after which the ledger ceases, but it is apparent from this that the mill was kept in constant repair and good order, and that improvements were made from time to time to increase its efficiency.

In 1843 Ashby made an agreement with a builder named John Muggeridge to let the eastern half of his property for building development, and seven houses fronting onto Cornwall Road (now Blenheim Gardens) were constructed.[B37] This agreement was to prove the beginning of the end of the windmill's working life, as it was due to the building up of the surrounding area that the mill eventually had to cease operation because of diminished wind power, and the milling process carried on elsewhere.

At some time between 1841 and 1845 the death occurred of John Ashby, the founder of the family business, and his will, which was signed on 27 December 1841 and proved in September 1845, directed that all his property (which was still in mortgage to Edward George) should be left to his son John, of the Lower Mill, Carshalton.[115] The will further directed that the beneficiary, in conjunction with Henry Dymond, a chemist of Croydon, should act as trustee for the property and should sell or lease the business in the Brixton mill to Aaron Ashby, or else to his younger brother Joshua in the event of Aaron's refusal of the business. Although the Rate Books for 1849 and 1850 assess Aaron for the house and mill, it seems that he decided not to continue in the milling business at Brixton, and after 1850 we hear no more of him.

42. Ashby's mill before restoration, showing the gallery still in position round the cap.

Further information about the Ashby family is provided by the Census of 1851,[116] which lists Hannah, aged 72 (not quite consistent with her age as given in the 1841 Census) as head of the household, giving her place of birth as Sevenoaks, Kent, and Joshua, her son, aged 30, now Master Miller employing four men, whose place of birth is recorded as 'Surrey, Brixton Hill'. Other persons listed as being in the Mill House at the time of the Census are John N. Allwin, apprentice miller, Anne Lutchford, house servant (both aged 16), Samuel Sholl (43), baker employing one man and one boy, and a visitor, Sarah Ashby, aged 46, born at Bidborough, Kent, whose profession is given as 'Miller's Daughter'. In the Mill Cottage, a separate building, was Jonathan Lutchford (born at Sevenoaks), miller, his wife Elizabeth, their son William, also a miller, and their six other children, three girls and three boys. As Hannah Ashby was also born at Sevenoaks, it seems likely that Lutchford was her maiden name, and that members of her family came over from Sevenoaks to assist in the running of the mill. Soon after this Census, Joshua Ashby married, and his son, also named Joshua, who was to carry on the family business after his father, was born in 1858.

During the 1850's the neighbourhood around Brixton Hill was rapidly built up over the open cornfields, and this proved to be too much for the efficient working of the mill which was thus deprived of its motive power. In 1862 the milling business was transferred to water-mills at Mitcham, and two years later the sails of the windmill were removed and the tower turned over to storage purposes.[B3] Some years after this a gallery was constructed round the exterior of the cap, from which a wonderful view of London could be obtained. At about this time the mill was sketched by an artist who 'replaced' the sails in his drawing, whilst retaining the gallery (*fig.43*),[117] although the two had never existed on the mill at the same time as there is no room at the front of the cap for both the gallery and the sails (*see fig.42*).

43. A nineteenth-century sketch of Ashby's mill, in which the artist has 'replaced' the sails for the sake of greater realism.

Commencing in November 1900 the *Brixton Free Press* began to carry advertisements on behalf of Joshua Ashby & Sons, whose address is given as Brixton Hill & Wandle Mills, Mitcham.[118] The firm is described as 'Millers, Corn & Forage Merchants' and the advertisements state that vans call on residents in the vicinity to deliver their products, including 'Fine wheat meal for brown bread'. This series of advertisements was intended to increase the trade in the Brixton area prior to closing down the Mitcham branch of the firm. In August 1902 the lease of the water-mill on the Wandle expired, and the firm gave the mill at Brixton a new lease of life by fitting it with a steam engine (later replaced by a gas engine) to drive the stones. Writing in 1964, an elderly inhabitant of Brixton who had lived opposite the mill as a child remembered that on the occasion of the coronation of King Edward VII in August 1902 the sails of the mill were turning decorated with fairy lamps, but this must have been some kind of makeshift arrangement as the real sails had been removed some thirty-eight years previously. The same informant said that her husband had been employed at the mill as a youth of 17, and at that time the manager was a Mr Baylis, with two brothers, Richard and Frederick Brains, also working there.[119]

Having leased the premises for almost a hundred years, in 1912 the Ashby family were able to obtain the freehold of the land, mill, and associated buildings. By this time the property had passed into the ownership of the Middlesex Hospital, and the freehold was transferred from them to Miss Hannah Ashby Ashby, Bernard Gideon Ashby, and Joshua John Ashby on 25 July 1912, on payment of a sum of £2000.[120]

In 1913 there was a sudden spate of interest in the press, which had just discovered that there was a working windmill (albeit gas-powered) to be found in the central London area. The *Daily Express* interviewed the manager, his name here spelt as Mr Bayliff, who declared that many people living in the locality could remember the time when they used to glean in the wheat fields and take their gatherings to the mill to be ground into flour, and that among his customers were many who had bought flour from the firm in the days when the sails of the mill were at work. He also said that stone-ground flour was declining in popularity as most people preferred the white flour produced by the roller-mill process.[121] A short paragraph in the *Pall Mall Gazette* expressed the hope that the extensive building of red-brick villas in Cornwall Road would not involve the destruction of the mill, and thus drew attention to the ever-present danger to the windmill from the activities of property 'developers'.[122] A leaflet produced by the firm at about this time gives a list of prices for various products, and gives prominence to the fact that the flour and wheat meal were produced by stone grinding, which was claimed to be beneficial to the health(*fig.47*)[123]

In the days when the Ashbys were leaseholders of the mill and associated premises, documents were drawn up from time to time, transferring the lease

44. A general view of the mill yard, now laid out as a public garden.

to various members of the family. This practice continued after 1912 in respect of the freehold, and a deed of 1927 is of particular interest in that it has an attached schedule of the machinery contained in the mill at that time. It reads:

One 10″ Chaff Cutter No.63 Hand Power
One Nominal 8 H.P. National Gas Engine complete with Water Tank and piping, exhaust piping and exhaust Box, Gas supply piping, Belt Guards, wood bench and Engine Room Fittings
One 2⅜″ Main Driving Shaft complete with Coupling 2 Wall Boxes and bearings, 2 brackets and bearings and 16 Driving Pulleys
One 6″ Single Leather Driving Belt 37 ft
One Oat Crusher with 2 Rollers 36″ × 5″ with F & L Pulleys and striking gear and complete with one delivery elevator 6″ × 3″
One Elevator 6″ × 3″ approximately 50 ft long with feed and delivery Spouts at each mill Floor
One 3 ft Millstone Mill with Iron Hursting with Wood Mortice and iron bevel gearing and driving countershaft, fitted with hand crane for lifting Millstones
One No. 1 Horizontal Grain Scourer and combined seed aspirator with Dust Truck to Mill. Two feed shoots and slides and 8 spare sifting screens.
One Flour Sifter and dresser with revolving Drum in varnished deal case with feed and delivery shoots and one sliding tray flour feeder with countershaft fitted with eccentric drive
One Self Raising Flour Mixing Machine with Feed Bin and delivery shoots
One No. 1 Standard D. Centrifugal Scalper fitted with feed, delivery and tails shoots mounted on and combined with One D. Centrifugal Dressing Machine complete with delivery shoots
One 4″ Worm Conveyor 12 ft in Varnished Deal Case and One 10 ft Elevator 3″ × 6″ with delivery and By pass shoots
One Single 3 cwt Sack Hoist with Shaft, Bearings, Belt Pulley, and lever tightening gear, hand ropes controls, Cat Head, 60 ft Chain and Hook Also 2 countershafts for driving same carried through Mill Floors
One Wood Exhaust Fan (requires repairs) and trunk inlet and exhaust piping ducts and also include all machine and belt timber guards
One Pair Milling Rolls 12″ × 9″ (not been used for some time) ?
One Grindstone 18″ fitted in C.I. frame
One 16″ Single Cake Mill.

Whether all these items were actually housed in the mill itself or whether some of them were installed in the various outbuildings is not made quite clear, but the list is an interesting inventory of the equipment in use after the mill had been converted from its original wind-driven mode of operation.[124]

Writing in about 1930, R. Thurston Hopkins stated that 'The old landmark stands amid rural surroundings of gardens and old-fashioned courtyards, forming a striking picture at night, when its round windows twinkle, and its tower rises black and sharp and dramatic against the sombre sky.'[B13] Despite the fact that the windows are square rather than round, and the surroundings distinctly less than rural, this description catches the atmosphere of the mill and its surroundings at dusk, but despite its attractive appearance its days as a working mill were numbered. By 1932 the gallery round the cap had become unsafe and was disused, and although an article in the *Brixton Free Press* in July 1933 celebrated the continuing functioning of the mill, claiming that there was an increased demand for the wholemeal flour, it finally ceased operation on the following year.

In May 1934 the Lambeth Borough Council approached the Society for the Protection of Ancient Buildings with a view to investigating some kind of joint action to secure the preservation of the mill now that it had ceased actual work. Accordingly, in June 1934 the Honorary Technical Adviser to the Society visited the mill and made the following report:[125]

> This mill was formerly used by Joshua Ashby & Sons – millers – who were established in 1816, and the mill has every appearance of having been built about that time. It is a brick built tower mill with a wooden boat-shaped cap from which the sails and winding gear have been removed.
> It stands well back from the road, access being obtained by a small carriage way running past the mill cottage, and there is a fairly large yard in front of the mill.
> All the old machinery has been removed from the mill and modern machinery installed, but the mill building is in excellent condition. The following repairs appear to be necessary:
> The cap needs small repairs to the weatherboarding at the rear where it has rotted. Dustfloor needs new window or hatch 12″ × 12″; one new light in existing window 8″ × 6″.
> Sack trap needs new leather hinges.
> The Bin floor requires a new hatch same size as the above and the sack trap needs repair.
> Dustfloor needs new window or hatch 12″ × 22″; one new light in existing window 8″ × 8″.
> Second Floor – 2 new windows or hatches required 24″ × 24″ and 2 new lights in existing windows 10″ × 11″.
> Stone floor – slight repairs to the floor.
> There is also a stage round the outside of the cap which has been condemned as unsafe, but without the aid of a long ladder it is not possible to examine this properly.
> Apart from the stage the whole mill building can be thoroughly cleaned and put into first rate order for the sum of about £25.–.–. TWENTY FIVE POUNDS.
> In addition to the above there is a range of outbuildings built around the base of the mill extending up to the level of the first floor. These are very extensive and appear to be in very fair order. The whole place would make a most excellent unemployment centre or club premises, and I should like to suggest that this property be preserved with the idea of using it to this end.
> The mill is of great interest as being almost unique in London, and is quite typical of the traditional practice of windmills built at the time of its erection. It would be a great pity if such a landmark of Old Lambeth was destroyed.

On Sunday, 5 May 1935, Joshua Ashby died at the age of 77. He was the grandson of the original owner of the windmill, which had remained in the hands of the same family firm for its entire working life. He had become something of a recluse towards the end of his life, and although he was a Quaker he did not actively practice his religion. His hobby was botany – he was a Fellow of the Royal Botanical Society – and he kept an interesting garden at the rear of the mill house where he grew rare plants, shrubs and flowers.[126] With his death this branch of the Ashby family came to an end, and the mill and surrounding property were left to his housekeeper, Miss Marshall, to be administered by a Trust.

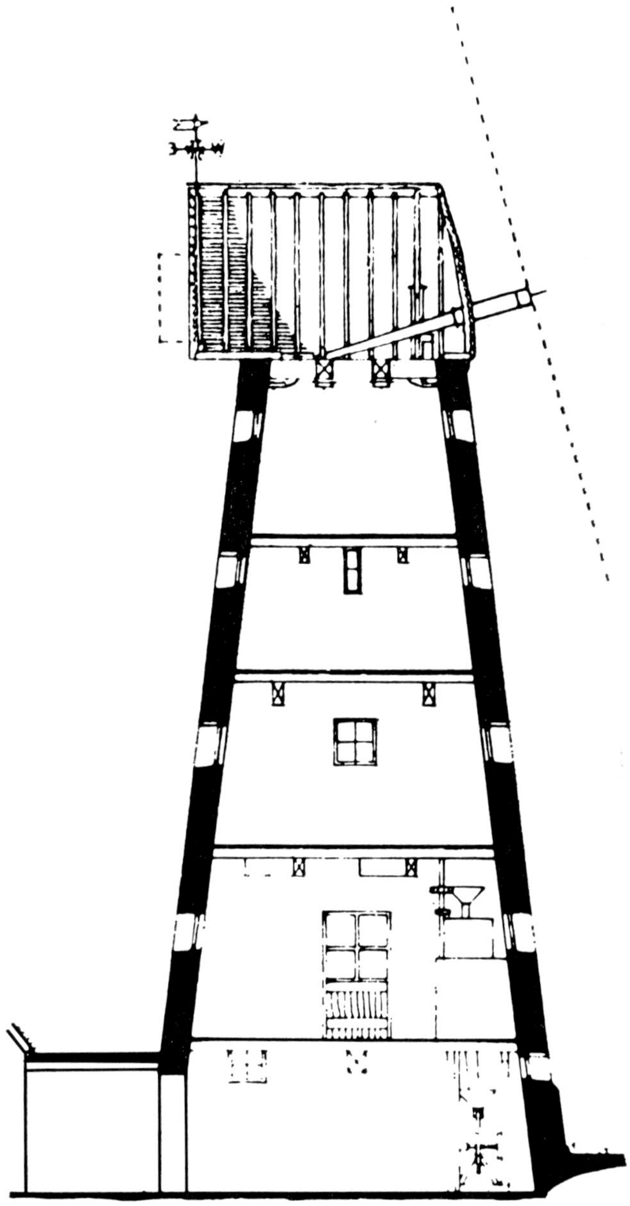

45. A cross-section of the mill before restoration.

The following year there was another rash of newspaper articles about the mill, this time due to the activities of a certain Mr John Ward, who visited the offices of the *Brixton Free Press* in September 1936 to announce that he had bought the mill on behalf of a milling syndicate, and intended to fit new sails and have the mill working again before Christmas. The story was taken up by various local and national newspapers, but it was discovered to be a hoax after denials by the owners that any sale had taken place. Nothing further was heard from Mr Ward, who had claimed to be the last windmiller in London.[127]

As a result of this publicity, the Society for the Protection of Ancient Buildings took up with the owners the question of the preservation of the windmill in its original form, but received the reply that as it was only built in 1816 the Trustees did not consider it to be an ancient monument and therefore were not prepared to sustain the expense of maintenance of the structure. Both the Lambeth Borough Council and the London County Council were, however, interested in preserving the mill as a building of historical importance, but it was felt that the price of £9500 asked by the Trustees for the site and buildings was too high, compared with the value of £6000 as estimated by the L.C.C.'s surveyor, and then the Second World War intervened to prevent any further negotiations from taking place.[128]

After the war, the mill itself was still in reasonable condition, although some of the outbuildings had apparently been slightly damaged by bombs which fell about 50 or 60 ft away, according to the owners. A book published in 1947 claimed that the mill was still working by means of the gas engine, but this was quite erroneous, as it is clear that all the buildings except the Mill house were virtually derelict at this time.[B2] In June 1950 the condition of the damaged outbuildings had deteriorated so much that a Dangerous Structure Notice was served on the owners by the L.C.C. District Surveyor, and in August the following year the War Damage Commission approved the owner's application for a grant to carry out the necessary repairs and suggested that the restoration should be carried out in consultation with the Ancient Monuments Division of the Ministry of Works. In September 1952 the District Surveyor found that the terms of the Dangerous Structure Notice had been complied with and that the buildings were now safe.

At this time interest again centred on the mill as a historic building, and various attempts were made to secure its preservation. In January 1951 the Historic Buildings Council of the Ministry of Housing and Local Government listed the mill as a Grade II building, and in the same month Mr Alan Reid reported to the Society for the Protection of Ancient Buildings that he had visited the mill and found the tower to be in good condition, and also the cap, except for the gallery which was defective and dangerous. The floors were all sound except for the ground floor which needed attention, but the outbuildings were in poor condition. At the suggestion of the Lambeth Civic Society, the Lambeth Borough Council was again exploring the possibility of purchasing the mill and laying out gardens round it, and to this end meetings were arranged with the owner's solicitors and the S.P.A.B., but once again the price of £10,000 including painting the mill proved to be the main stumbling block. In the course of the discussions the owner's solicitor stated that it was the wish of the late Joshua Ashby that the mill should be preserved as a relic of bygone days in Brixton.[129] As a result of these meetings, in September 1951 the Lambeth Borough Council decided in principle to buy the mill, subject to an approach being made to the L.C.C. for financial assistance towards the cost of purchase, and there the matter rested for the time being.[130] The cost of

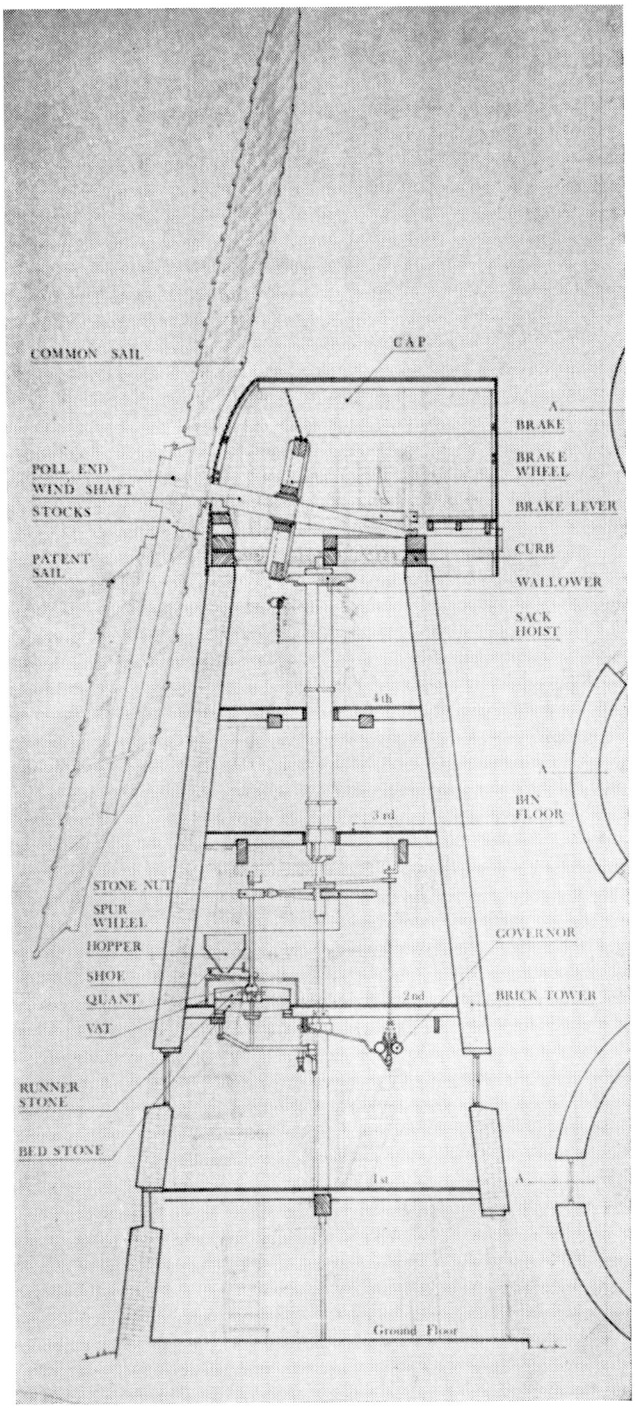

46. A cross-section after restoration, showing the sails and internal machinery.

restoring the mill to a reasonable condition was estimated at about £1500 at this time.[131]

Three years later came another upsurge of press reports concerning the mill. It appeared that the L.C.C. had received an application from a property developer who wished to demolish the mill and erect blocks of three-storey flats,[132] and although the application was turned down on the grounds that the site had been zoned as an open space in the County of London Development Plan of 1943, the whole question of the security of the mill's future was brought into the open once again. The Lambeth Borough Council reconsidered the matter, but finally decided that it would be too costly to acquire the site, as although the owners were now asking a sum of only £4800, the cost of renovation and laying out gardens was now

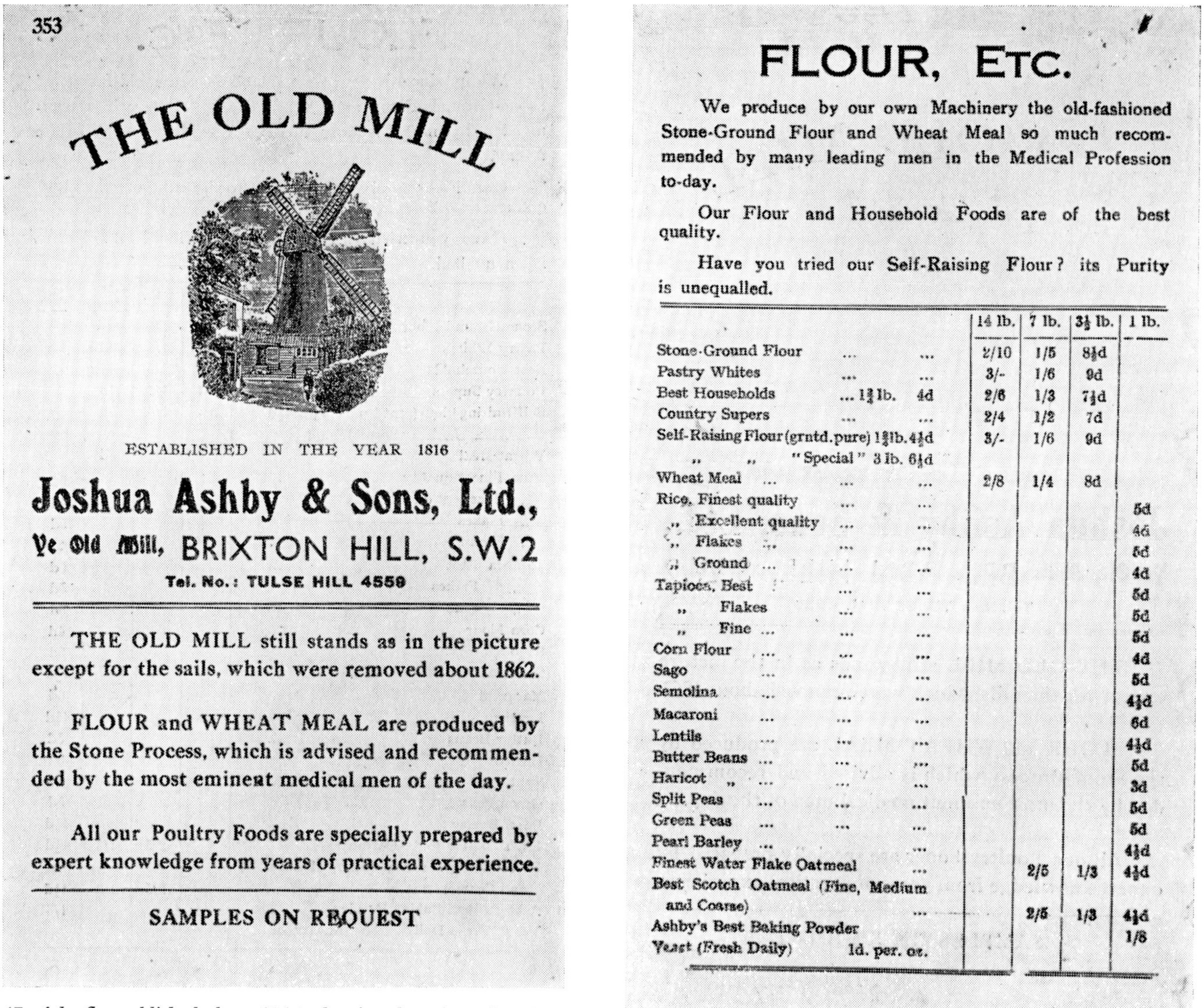

353

THE OLD MILL

ESTABLISHED IN THE YEAR 1816

Joshua Ashby & Sons, Ltd.,

Ye Old Mill, BRIXTON HILL, S.W.2

Tel. No.: TULSE HILL 4559

THE OLD MILL still stands as in the picture except for the sails, which were removed about 1862.

FLOUR and WHEAT MEAL are produced by the Stone Process, which is advised and recommended by the most eminent medical men of the day.

All our Poultry Foods are specially prepared by expert knowledge from years of practical experience.

SAMPLES ON REQUEST

FLOUR, ETC.

We produce by our own Machinery the old-fashioned Stone-Ground Flour and Wheat Meal so much recommended by many leading men in the Medical Profession to-day.

Our Flour and Household Foods are of the best quality.

Have you tried our Self-Raising Flour? its Purity is unequalled.

	14 lb.	7 lb.	3½ lb.	1 lb.
Stone-Ground Flour	2/10	1/5	8½d	
Pastry Whites	3/-	1/6	9d	
Best Households ... 1¾ lb. 4d	2/6	1/3	7½d	
Country Supers	2/4	1/2	7d	
Self-Raising Flour (grntd. pure) 1¾lb. 4½d	3/-	1/6	9d	
„ „ "Special" 3 lb. 6½d				
Wheat Meal	2/8	1/4	8d	
Rice, Finest quality				5d
„ Excellent quality				4d
„ Flakes				5d
„ Ground				4d
Tapioca, Best				5d
„ Flakes				5d
„ Fine				5d
Corn Flour				4d
Sago				5d
Semolina				4½d
Macaroni				6d
Lentils				4½d
Butter Beans				5d
Haricot „				3d
Split Peas				5d
Green Peas				5d
Pearl Barley				4½d
Finest Water Flake Oatmeal ...		2/5	1/3	4½d
Best Scotch Oatmeal (Fine, Medium and Coarse) ...		2/5	1/3	4½d
Ashby's Best Baking Powder				1/8
Yeast (Fresh Daily) 1d. per. oz.				

47. A leaflet published about 1914, showing the prices of various products supplied by Joshua Ashby & Sons Ltd.

estimated at £12,000.[133] Eventually in 1957 the L.C.C. purchased the site, and work was commenced immediately on laying out the gardens; about ¾ acre was to be made ready for immediate use, and eventually the gardens were to be extended to the whole of the 2½ acres.[134]

In 1960 the outbuildings were demolished, and in 1963 Mr R. Thompson, a millwright from Alford, Lincolnshire, commenced work on the restoration of the mill itself on the basis of a comprehensive report on the condition of the mill prepared for the L.C.C. in 1958 by Mr Rex Wailes.[135] At first it was hoped that the mill could be made to work again, but this was found to be impossible, and the restoration was confined to bringing the mill as near as possible to its original appearance. Some of the gas-driven machinery was removed, and wind-driven machinery installed, most of which was brought from a derelict tower mill at Burgh-le-Marsh in Lincolnshire, which was then being dismantled, and the rest was newly made. The sails were constructed in accordance with those shown in a photograph of the mill taken before 1864, when the original sails were removed (*fig. 48*). At Easter 1968, the interior of the mill was opened to the public for the first time in its newly restored state.

The mill as it stands today is a good example of the state of windmill technology at the time of its erection (*fig.49*). It is a tower of five storeys, tapering from an external diameter of 22 ft at the base to 12 ft at the top. Its height is 39½ ft to the top of the tower, the curb and cap adding a further 9½ ft, making 49 ft in all. The walls are 1 ft 6 in thick, and the brickwork is coated with tar on the exterior to afford some protection against the weather. Although the mill itself was built in 1816, many timbers used in its construction are apparently older than this and are reputed to be old ship's timbers.

The cap is boat-shaped, and rests on a 'dead' curb of metal around the top of the brick tower (*fig.50*). Six centring wheels keep the cap in position horizontally (*fig.51*), and there is also a rack running round the interior of the top of the tower to enable the cap to be turned from the ground by means of an endless chain passing over a wheel on the exterior of the cap (*fig.52*). During restoration work it was found that the curb at the top of the tower was not exactly circular, thus preventing the cap from being turned round as was originally intended in the restoration plan.[136] The gallery, which was fitted round the cap after the sails had been removed in 1864 and had subsequently become unsafe, has now been removed to enable a set of sails to be fitted to the mill.

48. A photograph of Ashby's mill taken before the original sails were removed in 1864.

49. Ashby's mill as it appears today.

There are two pairs of sails, fixed to the windshaft through a canister or poll-end (*fig.53*), one consisting of common sails, and the other of the patent type; this arrangement being in accordance with the old photograph of the mill which shows the two types of sail in use simultaneously (*fig.48*). The patent sails were originally operated by means of a striking rod passing through the centre of the cast iron windshaft, to emerge at the rear of the cap where a wheel and endless chain mechanism enabled control to be exercised from a stage which encircled the mill at first floor level. This stage may be seen in an early pencil sketch (*fig.54*) and would have been used primarily for adjusting the canvas on the pair of common sails. This sketch also shows a long pole reaching down to the stage level, to enable the cap to be turned easily in the event of failure of the rack mechanism. A similar system was in use at Randall's mill, Nine Elms (Site 6). Access to the stage was provided by a door in the body of the mill. The striking rod mechanism has not been restored, however, and the sails have been left without shutters or canvas in order to reduce wind resistance.

Inside the cap, the windshaft carries the large brake wheel (which has been immobilised by the insertion of wooden blocks to prevent the sails turning)

50. Rear view of the cap.

51. One of the six centring wheels which keep the cap in position.

(*fig.55*) and this engages with the 'wallower' gear wheel at the top of the main vertical timber drive shaft. Just below the wallower is the sack hoist; a wooden drum which could be made to turn by means of a friction drive from the wallower (*figs.56* and *57*). By moving a lever, a wooden wheel attached to the drum could be moved to act on the underside of the wallower, thus enabling sacks to be hauled up to the top of the mill through trapdoors in each floor (*fig.58*). The storage bins which would have originally stood on the third floor to contain wheat waiting to be ground are now missing.

The millstones are situated on the second floor, and are surrounded by a wooden casing known as the vat, which supports a wooden hopper through which the grain flowed on its way to the stones (*fig.59*). There were originally two pairs of stones, but only one pair has been restored in order to leave sufficient space for visitors to inspect the mechanism.[137] A large gearwheel (the 'great spur wheel') on the main shaft acted on a small gear ('stone nut') and spindle attached to the upper, or runner stone of each pair, and these could be engaged or disengaged at will (*fig.61*) After passing between the stones, the grain fell through a spout and was collected in sacks on the floor below. Also on the second floor is a hopper and

52. The rack mechanism which enabled the cap to be turned into the wind.

53. The method of attaching the sails to the windshaft by means of a canister, or poll-end.

set of stones dating from the time when the mill was operated by means of a gas engine (*fig.62*), and below this floor is a centrifugal governor which enabled compensation to be made for the increase in separation of the stones which took place as the speed of rotation increased (*fig.64*). On the first floor there would probably have been a wire machine or bolter to separate the flour from the bran, and the ground floor space would have been used for storage of the finished products, awaiting sale or use in baking bread.

Nothing now remains of the original outbuildings which stood round the mill (*fig.65*) and included the Mill House, the Mill Cottage, and various buildings associated with the bakery which operated in conjunction with the mill, but the whole area has been laid out as a pleasant garden with a children's playground and forms a welcome addition to the amenities available to the residents of this otherwise drab area.

55. The brake wheel, which engages with the main vertical shaft.

54. An early pencil sketch of Ashby's mill, showing the stage at first floor level and the adjacent brick-fields.

56. The sack hoist, with its friction drive wheel.

57. Another view of the sack hoist.

58. The lever which operated the sack hoist.

59. The stone floor, showing the great spur wheel, the hopper, and the vat enclosing the stones.

60. The shoe which directed the flow of grain into the eye of the runner stone.

61. The stone nut, disengaged from the great spur wheel.

62. The hopper and set of stones dating from the days when the mill was gas-driven.

63. The underside of the nether stone, showing the bearing for the stone spindle.

64. The centrifugal governor which controlled the distance between the stones.

65. A photograph taken before restoration work commenced, showing the mill house (now demolished).

Site 9: Morrish Road (formerly Mill Lane), Brixton Hill

About a third of a mile south of Ashby's mill stood another windmill, on the north side of a lane off Brixton Hill, now known as Morrish Road.[138] On 19 October 1799 Thomas Potter of Brixton Causeway took out an insurance with the Sun Insurance Company on a

> Wind Corn Mill only situate as aforesaid. No kiln therein. Timber. Not exceeding One Thousand pounds. Standing and Going Gears. Mill Stones. Wire Machines & Dressing Mills therein only not exceeding two hundred pounds.
> Stock and utensils including boulting cloths and other moveable utensils therein only not exceeding three hundred pounds.[173]

The mill first appears on a map of South London by James Edwards dated 1800,[B20] and three years later it was sketched by G. Shepheard, who showed it as a smock-mill with a waggon-shaped cap and with winding ropes or chains hanging down to a stage round the mill body.[139] In 1808 a picturesque representation of the mill was included in the Perambulation of the Parish of Lambeth,[140] but as in the case of Randall's mill at Nine Elms it is doubtful whether this was intended to be an accurate representation of its appearance (*fig.66*). It is drawn as a tall eight-sided smock-mill, with four common sails and a stage with a door in the mill body giving access to it. It has a dome-shaped cap and there is a door and window in the body below the stage. The first storey is apparently of brick or stone, the rest being of wood. The Perambulation describes the site as being 'Revd. Mr Fieldings house, windmill and land', but no mention of anyone of this name can be found in the Rate Books of the period.

Another view of the mill, dating from 1818,[141] shows it as a six or eight-sided smock, with Bleak Hall Farm in the background, but the representation of the mill is so simplified as to prevent any further information being gleaned from it (*fig.67*).

The first mention of the mill in the Lambeth Rate Books occurs in 1813, when 'Gostling' was rated for a house and windmill, the rateable value of the latter being £28. This entry is repeated in the following two years, after which there is a period of silence. Pigot's directory for 1823–4[142] gives two millers at Brixton; John Ashby and John Cushion, and the 1824 Poor Rate book confirms Cushion as being at the southerly mill, being listed under 'Brixton Causeway – Blicke Hall'. John Cushion remained at the mill at least until 1830, when the Rate Book entry bore his name, but the next time the mill is referred to, in 1839, the occupier is given as John

Paddy. As will be seen later, a miller of this name figured in the history of the windmill at Westow Hill (Site 12), and in all probability he came to Brixton Hill as a refugee from the events at Norwood. In the Census of 1841,[143] Paddy is listed as aged 35, profession miller, and place of birth outside the county of Surrey. His wife Anne and their six children are also listed.

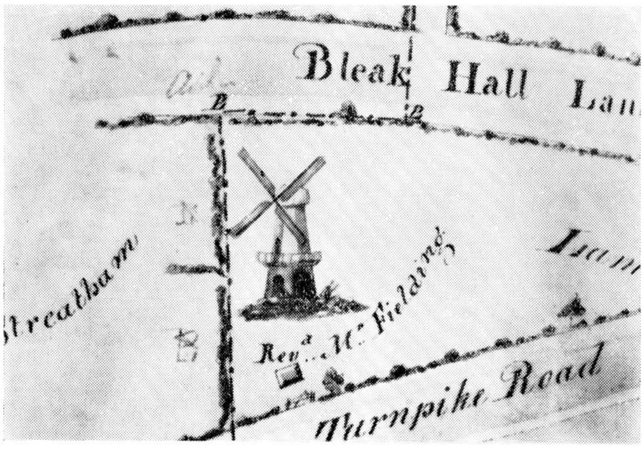

66. The windmill at site 9 (Brixton Hill), as shown in the Perambulation of the Parish of Lambeth, 1808.

Throughout the first half of the nineteenth century, the mill appears on various maps, the last two being those of Davis, dated 1844, and Wyld, 1861. However, as this last map claims to be based on the Ordnance Survey (of *c.*1816), it is likely that the mill disappeared soon after 1844. In the Rate Book for 1840, Paddy is assessed for the 'Upper Mill', but by 1849 the entry has become simply 'House & premises', and presumably the mill had been demolished during the interim period, as there is no further mention of it. The 1851 Census records a John Paddy aged 25, born in Kent (presumably a cousin or other close relation of the elder Paddy) who is unmarried and a Master Baker by trade, employing John R. Brunt, aged 40, Baker, and Thomas Harding, aged 17, an errand boy.

In 1858 an engraving was published entitled 'Sunrise in Autumn – View near Brixton', based on a painting by W. Collingwood Smith,[144] a watercolour artist who lived on Brixton Hill (*fig.68*). The view shows a tower-mill, apparently of only two storeys, with four shuttered sails and a waggon-shaped cap. A short distance from the ground there is a stage supported by struts, with a ladder reaching to ground level, and a door gives access to the mill body from the stage. A pole projects from the mill body and from this ropes hang down to the stage. Presumably this was some kind of brake control mechanism similar to that employed on some Dutch windmills, but it is extraordinary to find the pole projecting from the mill body instead of from the cap, as it would have been certainly struck by the sails when the cap turned. However, the cap itself does not appear to be capable of turning as it is supported by two external

67. Detail from a watercolour by G. Harley, dated 1818 and entitled 'Near Brixton Hill—windy day'.

struts from the mill body. In the side of the cap there is a small window in a gable, and at the rear there is a staging which would have once carried a fantail. Perhaps in the mill's last years the winding mechanism had become faulty and the cap had been permanently set to face the direction of the prevailing winds at the top of Brixton Hill. The sails are of the double-sided variety, and the windshaft appears to project a long way in front of the sails, which implies that the sails must have been spring sails rather than patent sails, as it would have been impossible to fix a spider mechanism to such a projecting windshaft. The date of the original watercolour cannot be ascertained, but the mill may have been out of use at the time the painting was made, judging by its generally run-down appearance.

No trace at all remains of the mill today, but its site is discernible from several maps, including the Parish map of 1824,[145] where it is shown as being on the north side of the lane, approximately two-thirds of the way along from Brixton Hill, and the Tithe map of 1843[146] shows it as octagonal in plan, thus confirming its type as a smock-mill.

68. 'Sunrise in Autumn—view near Brixton': an engraving from a watercolour by W. Collingwood Smith, dated 1858.

Site 10: Knight's Hill

Although the main road running through West Norwood is today called Knight's Hill, it was originally known as Knight's Hill Road, and the hill from which it took its name lies between the present Norwood Road and Rosendale Road, being bounded on the south by Lovelace and Thurlow Park Roads. A windmill stood on this hill in the first part of the nineteenth century, but apart from its appearance on several maps, there is no documentary evidence to show what it was like or what became of it.

During the middle ages there was a settlement here, and although there are references to mills in deeds and leases,[147] these appear as part of standard legal phraseology, and would apply to all other types of mill besides windmills. In the eighteenth century the land formed a detached portion of the manor of Leigham Court and of the parish of Streatham, and between 1792 and 1795 Lord Thurlow built a large house, designed by Henry Holland, at the southern end of the hill. Views of the house made in 1795 and 1805 show no sign of a windmill on the hill, which is depicted as treeless and barren, with a footpath leading to the top.[148]

In 1806 Lord Thurlow died, and his will instructed the Trustees to sell his land. An act of Parliament of 1809 empowered them to demolish the house, sell or lease the land, and to make roads, and the estate was dispersed between the years 1810, when the mansion was demolished, and 1846.[B37] The first map to mark the windmill is the first edition of the one-inch Ordnance Survey map (*c.*1816),[149] and this suggests that the mill was erected as a result of the development of Lord Thurlow's estates after his death. Several subsequent maps show the mill, including those of Bryant (1823), Greenwood (1829), and Davis (1841).[150] The last map on which it appears is that of Wyld, dated 1861,[151] and another version of the same map shows the proposed railway routes and the tunnel through Knight's Hill from Tulse Hill station. From these maps it appears that the mill stood on the very summit of the hill, directly above the position of the southerly mouth of the present railway tunnel.[152] The railway line from Tulse Hill to Peckham was officially opened on 1 October 1868,[153] and it is possible that the windmill was yet another victim of railway expansion, although it is not marked on a map of Streatham of 1843,[154] nor on the Tithe map of 1840,[155] where the land is shown as being owned by the Executors of Lord Thurlow, occupied by one Martin Samuel, and is described as 'meadow'. Allowing for the tendency of map-makers to publish reprints without an additional survey, it is likely that the mill was short-lived and disappeared sometime during the 1830's.

Today the top of the hill still has the appearance of open countryside, apart from some allotments on the north-eastern slope, and it seems that it has never been completely built upon. No traces of the mill remain, but the summit is a very exposed position and it is possible that the mill was blown down during a gale, never to be re-erected.

Site 11: Crown Point

Although some authorities have doubted the existence of this windmill on account of the inaccuracies of cartographers, it seems likely that a mill stood somewhere near Crown Point (i.e. the junction of the present Knight's Hill with Crown Lane and Crown Dale) during the sixteenth and seventeenth centuries.[156]

Writing in Volume 2 of *English Windmills*,[B9] Donald Smith confuses the mill symbol on the 1610 map of Surrey by John Speed[157] with a mill site at Brixton, on the grounds that the symbol appears under the word BRIXTON. However, inspection of the map reveals that the mill symbol appears amidst a complete phrase, reading BRIXTON OR ALLINGTON HUNDRED, referring to the whole area of Surrey north of the Hundred boundary, and in no way connected with the village of Brixton itself (*fig.69*). The earliest map to show the mill is that of Norden, dated 1594,[158] which places it on a stylized hill to the south of a similar hill labelled Knight's Hill (i.e. the Knight's Hill referred to in the last chapter, and not the Knight's Hill of today, which was formerly known as Knight's Hill Road), and uses the conventional post-mill symbol of the time. Other maps follow, notably those of Whitewell (1604), Speed (1610), and Blome (1673),[159] but these all give substantially the same information, and would indeed appear to be copied from each other, or else from the original by Norden.

An anonymous map published about 1665 again shows the mill symbol,[160] but also includes the main north-south road (i.e. the present Knight's Hill) which is absent from the earlier maps. The mill is shown on the east side of the road, and, as in the other maps, to the north of the Hundred boundary, which has remained unchanged since Norman times.[B38] Further maps, by Saxton/Lea (*c.*1690),[B34] Bickham (1706),[161] and Lea (1706),[162] also show the mill, and although the presence of the symbol on the early maps is not given much credence by Farries and Mason, they do, however, credit Aubrey's map published in 1719 with a measure of reliability, and on this map the Crown Point mill also appears. A further confirmation of the accuracy of this map is provided by its indication of the windmill at Walworth, whose site can be verified from later maps of some reliability. The last map on which the mill appears is that of John Seller, dated 1733,[163] and there are no further documentary records relating to it. The evidence of these maps is insufficient to pin-point the exact site of the mill, due to the type of pictorial representation employed, and we can only say that it must have stood somewhere in the Crown Point area, possibly just to the east of the main north-south road.

There are no pictorial records of the mill apart from the conventionalized symbols used on maps, but we may assume it to have been a small two-roomed post-mill with a straight-pitched roof, similar to the mill on Clapham Common described above (Site 7), and that it was probably engaged in grinding corn rather than in one of the industrial manufacturing processes which were to be found nearer the centre of London.

69. Detail from John Speed's map of Surrey, published in 1610.

Site 12: Westow Hill

Little is known of the type or appearance of this windmill, which stood a short distance outside the Lambeth parish boundary at Westow Hill, but fortunately a large amount of information concerning the millers is preserved in the Croydon Rate Books,[164] which record an ever-changing succession of occupants.

The mill first appears in the Rate Books for the year 1813, when Anthony Holloway was assessed for a house and windmill, at a rateable value of £20. This entry appears in subsequent books until May 1815, when the rate was charged to Joseph Fletcher, and in August 1816 the occupant had become George Brown, who remained until the early part of 1818. Some interesting information regarding the end of Brown's tenancy has been discovered by Farries and Mason, and is recounted in their *Windmills of Surrey and Inner London*:[B19]

> George Brown, miller, came from Battersea in 1816 and occupied the premises, which comprised, in addition to the windmill, a brick dwelling house, stables, pigsties, barns, an oven and other buildings. The owner, Anthony Holloway, described as a farmer of Woodside, Croydon, demised the premises to Brown at rack rent. On the 7th March, 1818, three-quarters of a year's rent was in arrears. Holloway, having discovered that his tenant had deserted the premises and had left them empty of goods, so that no recompense could be found in lieu of rent, called without delay upon Thomas Fish and David King, justices of the peace, and requested them to take action on his behalf. Forthwith the three 'did personally go to and view the said premises, and then and there, upon our own proper view, did find the said Complaint to be true, and did then and there affix, in the most notorious parts of the said premises, to wit on the outer Doors of the said Windmill and of the said Messuage or Dwelling House, a notice in writing under our Hands and Seals, and we the said justices on the twenty-third day of March then Instant, and now last, would return to take a Second view thereof.' On the 23rd neither George Brown nor a representative appeared, and, accordingly, Holloway was given immediate possession.

The next occupier was one Robert Cooper, whose tenancy was rather short-lived. His name appears only once in the Rate Books, in November 1819, and it has been crossed out and the note 'run away' written in pencil beside the entry. On the facing blotting paper is written 'Noakes for Mill etc.' and in April 1820 we find a full entry for Noakes, assessing him for 'House, Garden, Mill etc.' at a value of £21. However, this entry has also been crossed out in pencil, and 'V. Marlow for Mill etc.' written underneath. It will be remembered that a James Noakes made an appearance at the windmill in Mill Street, Lambeth Walk, sometime between 1820 and 1823, and it may well be that this was the same miller who had abandoned his tenancy at the Westow Hill mill.

In 1821 a full entry for Valentine Marlow appears, and this continues until August 1824 when the name 'Valentine' is crossed out in pencil, and 'Mrs.' inserted above. Whether Valentine Marlow had died or merely absconded is not clear, and in November 1824 John Marlow is assessed for the mill. Apart from a spurious reference to a certain Robert Smith in occupation in December 1824, John Marlow continued in tenancy until 1826, when the entry is annotated in pencil: 'gone away the latter end of June, and has not been seen since'. Later in the year the Rate Books list V. Marlow as being assessed for the mill, but no rates seem to have been actually paid, so perhaps this was merely a nominal entry in the absence of any other tenant, and did not necessarily mean that Valentine Marlow was back in charge of the premises. The following year we find first the name of George Heffar, followed in August by that of Z. Spice, both being rated for 'House and Windmill', whereas in October Zachr. Spice is rated for the house only, and William Heffar assessed for the mill itself. During 1828 the entry for the mill is the same (the surname being spelt as Heffer), but in February 1829 it is listed as being empty, and remained so until October when it was taken on by Gustavus Wilmot. By the side of Wilmot's name is that of B. Digby, entered in pencil, and it is probable that these two were in partnership in running the mill.

In May 1830 the windmill is assessed to 'Gus. Wilmot or Benjamin Digby', and it seems that some dispute over the rate payments had arisen, for on the facing blotting paper is written the cryptic note 'Anthy. Holloway to pay – 23rd Apl., Wilmot –, H. T. Marlow – Michas., Goacher.' In July the entry merely records 'occupier', which in August becomes 'B. Digby or occupier'. By October a Thomas Goodman had become the tenant, and he continued until October the following year, when he relinquished his tenancy to H. Marlow, but at the end of the year the entry carries the note 'Mill damaged early in December and not used since'.

It seems that the damage was not made good until well into the following year, and it is not until November 1832 that a John Pecke (or Peche) is listed as occupier. However, this arrangement does not seem to have lasted for long, as in March 1833 we find 'Homewood' as tenant, and in April the mill is once again listed as being empty. In June 1834 Isaac Marsh is listed as the occupier, but the mill must presumably have been working again for some time prior to this, for against the entry is written in pencil 'John Paddy terminated'.

In the midst of all these comings and goings, one's sympathies extend to the unfortunate rate-collector, whose diligent efforts to keep the records up to date are evident from the Rate Books. In attempting to record the name of each miller for the appropriate date, he often becomes confused, and below the entry for July 1834 which lists Isaac Marsh, he writes: 'now Paddy, now Holloway, now Earl, now Holloway', and again in January 1835: 'John Paddy, now Holloway, now Jacob Earl', and finally gives up completely, simply writing 'occupier' in the June 1835 book.

Although John Paddy was listed in June 1834 as having 'terminated', his name appears frequently after this date, both alone and in association with the names mentioned above. In October 1835 we read 'occupier – Randall, now Paddy – Churcher from August', which is amplified in a note on the blotting paper: 'Randall and Jeans left Paddy and Churcher from August'. However, Randall was back by March 1836 when he was assessed for 'cottage and mill', but in July the same year C. Churcher took over the premises, a note stating that 'this March Rate to Randall who has left'.

Charles Churcher continues to be listed as tenant until January 1838, when he was assessed only for a cottage, the 'Mill only' being charged to 'Pocknell', qualified in July 1838 as 'Pocknell Jnr.' In June 1839 the windmill and yard were still charged to Pocknell, but the house was now occupied by Paddy, which is later amplified to Thomas Paddy. It is interesting to note that at some time before 1839 a John Paddy took up the tenancy of the windmill on Brixton Hill (Site 9), and this was doubtless the miller mentioned in the last paragraph, settling down to a peaceful existence elsewhere after the turmoil and uncertainty of life at Westow Hill.

In July 1840 the name of the tenant changes from J. Pocknell to W. S. Pocknell, and this continues for several years until August 1846, when Charles Churcher re-emerges as miller. The Census of 1841[165] lists William Pocknell, aged 20, miller, his wife Jane, aged 25, and their daughter Sarah, aged 10 months. In the same row of houses, Woodman Row, are listed Thomas W. Paddy, baker, aged 35, and his family, and William Pocknell, aged 50, described as being of independent means – presumably the Pocknell who also appears in the Rate Books as the owner of a number of houses in the vicinity. Another miller listed in this Census in Woodman Alley nearby is James Adams, aged 30, but whether he was connected with the windmill or with some other type of milling is not stated.

Although Anthony Holloway had remained the owner of the property throughout the changing tenancies, Charles Churcher seems to have purchased the mill and cottage in 1846, but his ownership could not have been very profitable, as by July 1849 the mill is listed as being empty, Churcher having leased it to one Saxby since the previous May. It remained empty and unused for the next few years, and in May 1853 the entry for the mill is crossed out and the assessments erased, and there is no further mention of it in the Rate Books after this date. Gray's Croydon Directory of 1851 and 1853[166] lists William S. Pocknell as Miller of Woodman Road, Westow Hill, but the 1855 edition lists him only as Cornmerchant and Grocer (at the same address), whereas the 1851 Census[167] makes no mention of Pocknell, but lists Thomas Paddy, baker, whose age has now apparently increased to 47 in the ten years since he was 35, and gives the further information that he was born at Farnborough, Kent.

Thus ends the story of the millers of the Westow Hill mill, which must be one of the most detailed lists of the occupiers of a windmill now extant. The most intriguing question about the whole business is why there should have been such a turnover of tenants. Perhaps, as Farries and Mason suggest, the working conditions were very uncomfortable or the rental was too high, or perhaps it was not possible to work the mill efficiently due to the wind power being cut off by the rise of neighbouring buildings during the development of the area. Although there seem to have been disputes between the landlord and tenants, most of the names mentioned also appear in the Rate Books as tenants or owners of houses in the vicinity, sometimes even after they had absconded from the windmill, thus showing that time eventually healed the differences between the parties. Apparently various people in the neighbourhood took it in turns to grapple with the problems of working the mill, and several of the names crop up quite frequently in connexion with another windmill which stood on Croydon Common, which is outside the scope of this survey.

The few maps which show the mill include that of Allen dated 1825[B25] and the Tithe map of 1844,[168] and from these its position can be ascertained as 80 ft south of Westow Hill, at the rear of what is now No.30 Westow Hill.[169] The Tithe Award of 1844 gives some detail of the premises,[170] listing Anthony Holloway as landowner, and the occupiers as Charles Levin (house, baker's shop, stable and garden), William Sawyer Pocknell (windmill, yard and garden), and Thomas William Paddy (house, baker's shop and yard).

Speaking in 1907, a Mr E. Wheeler, wheelwright, of Westow Hill, recalled:

> The windmill was at the top of Gipsy Hill by the side of the Windmill Tea Gardens. The windmill was kept by a miller whose name was Paddy, and who occupied the millhouse nearby, a part of the house is still existing next door to Mr. Squire's Printing Works (*fig. 70*). This windmill was stopped in the year 1853, and was pulled down soon after.[B32]

As regards the type or construction of the mill, there are no views or drawings now existing which depict it, but some information may be obtained from entries in the ledger of William Ashby, the millwright previously mentioned in connexion with Ashby's mill, Brixton.[171] At the end of 1836 Anthony Holloway engaged Ashby to effect some repairs and alterations to the mill, and work accordingly went ahead. It seems that Ashby had also done some work previously for Holloway and was experiencing some difficulty in obtaining payment, as his Day Book records letters sent to Holloway in June and November 1836 and February 1837, all asking for payment within two weeks of the outstanding sum of ten guineas. Holloway finally paid up on 6 June 1837, thus enabling work on the new modifications to begin.

70. Westow Hill: remains of the mill house, 1970.

On 16 November 1836, Ashby's son Thomas had gone to Norwood to take the 'dimentions' for a new pair of sails, and on the 23rd of the same month Ashby himself travelled to Greenwich to look out the timber required for the work, which he brought home with him. The timber for the middlings was 70 ft long, and pieces for the whips were 36 and 40 ft, respectively, being purchased from Dawsons and Co. at a total cost of £16–0–7, of which Ashby paid £5 cash on account. On 10 June 1837, immediately following Holloway's settlement of the previous bill, work commenced on fitting the new sails, which were of the patent type, containing a total of 124 shutters 3 ft 3 in. long by 12 in. wide, and fitted to the windshaft by means of an iron cross. At the same time the windshaft was drilled to take the striking rod which operated the shutter mechanism.

As indicated by the Rate Books, from January 1838 W. S. Pocknell, Jnr. took over the mill, and he seems to have begun his tenancy by putting things into good order. Ashby's ledger for 1838 lists several items of work put in hand for Pocknell, including new quant brasses, spindles, and wedges. The main item was a new pair of Peak stones and their associated machinery, which were installed in April of that year. Another item of interest is a charge for two men working at French stones (i.e. dressing them), which indicates that at least two pairs of stones were being driven in the mill at this time. In December, 3 ft of timber was used for a sail end, probably for repairs due to the effect of the winter gales, and in July the following year an oak wedge 'to hang Wollar with' (i.e. wallower) was supplied at a cost of six shillings.

No further work on the mill is recorded in the ledger, but it seems that Ashby again experienced some difficulty in obtaining payment for his work, for although Pocknell had paid £7 on account at the time of fitting the new sails, the balance was not forthcoming, and on 14 May 1840 Ashby made the entry: 'I wrote to W. S. Pocknell to send me a check or bill on short date – £10–6–5.'

Other names connected with the mill, such as John Paddy and Gustavus Wilmot, also appear in Ashby's ledger, but the dates do not appear to coincide with the periods of their tenancies at Westow Hill, and we must, therefore, assume that the entries relate to work carried out for these millers at other windmills in Ashby's extensive area of work.

In his booklet *Some Glimpses of Norwood*,[B35] Alan R. Warwick states that traces of the windmill were still visible as late as 1961. Apparently, when viewed from a window in the adjacent printing works, a circular brick foundation could be seen on the ground,[172] which would suggest that the mill was either a brick-built tower-mill, or else a post-mill with a brick roundhouse. However, when the site was visited by the present author in the summer of 1969, a close inspection of the ground failed to reveal any trace of this foundation, but the manager of the printing works confirmed the site of the mill, and said that the piece of ground was still known as Windmill Yard.

Glossary

Air poles: rods connecting the roller blinds to the spider, to enable them to be adjusted simultaneously.
Backs: larger versions of whips, used with a central iron cross to which they are bolted.
Backstays: wooden bars at the back of a sail to maintain the angle of twist.
Batter: slope of the sides of a smock-mill.
Bays: spaces between the sail bars.
Bed stone: the lower, fixed, millstone. (also Nether stone).
Bill: tool used for dressing millstones.
Bin floor: where the grain is stored before being fed to the hopper for grinding.
Blue stones: type of millstones imported from Cologne. (also Cullin stones).
Bolter: machine for separating the flour from the bran after grinding.
Brake wheel: large gear-wheel on the windshaft, driving the wallower. The brake acts on its rim.
Breast: upper part of the front of a post-mill. (also Head).
Breast beam: main front beam in a post-mill, which bears the windshaft. (also Weather beam).
Bridge: iron bar across the eye of the upper millstone to enable it to be fixed to the spindle.
Bridge tree: hinged beam carrying the lower stone spindle bearing.
Bridle irons: clamps used to fasten the backs of the sails to the arms of a central iron cross.
Buck: body of a post-mill (East Anglian term).
Burr stones: millstones built up from blocks imported from France: used for grinding wheat. (also French stones or French burrs).
Canister: cast-iron socket at the end of the windshaft, which holds the stocks. (also Poll-end).
Cant posts: the sloping corner timbers of a smock-mill.
Cap: top section of a smock or tower-mill, which can be turned to face the sails into the wind.
Clamps: timbers bolted to the sides of the stocks in order to strengthen them.
Collar: bearing at the bottom of a post-mill body to steady it against the post.
Common sails: consisting of wooden frames over which canvas is spread.
Composition stones: millstones made from a mixture of cement and carborundum or other abrasive.
Cross: a cast iron cross fitted to the end of the windshaft, to which the backs are directly bolted.
Cross trees: two horizontal timbers at right angles, which support the post of a post-mill by means of the quarter-bars.
Crown tree: the main beam of a post-mill, which revolves on top of the post.
Cullin stones: type of millstones imported from Cologne. (also Blue stones).
Curb: track on top of a smock or tower-mill wall, on which the cap turns.
Damsel: four-sided spindle which vibrates the feed shoe under the hopper.
Dead curb: where the cap slides on wooden blocks instead of running on rollers.
Double shuttered sails: in which shutters are fitted on both sides of the whips.
Dressing: the process of cutting the grooves on a millstone.
Driving side: the trailing part of a sail.
Dust floor: top floor of a tower or smock-mill.

Eye: hole at the centre of the upper millstone, which allows the grain to pass through.
Fanstage: framework fixed to the cap to support the fantail.
Fantail: windvane geared to keep the sails facing into the wind. (also Fly).
Fly: see Fantail.
French burrs: millstones built up from blocks imported from France: used for grinding wheat. (also Burr stones or French stones).
French stones: see French burrs.
Furrows: grooves cut into the surface of a millstone.
Gate: slide to regulate the flow of grain through the shoe.
Governor: centrifugal regulator to maintain a constant distance between the millstones regardless of speed.
Great spur wheel: large gearwheel on the upright shaft, which drives the stones through smaller gears.
Head sick: a post-mill leaning forward out of the vertical.
Hemlaths: long strips of wood at the outer edges of the sails.
Hopper: container which feeds grain to the stones via the shoe.
Horse: wooden framework which supports the hopper.
Lands: ridges between the furrows on the surface of a millstone.
Leading boards: wooden boards fixed to the leading edges of single-sided sails.
Listings: canvas straps controlling the blinds on roller-blind sails.
Live curb: where the cap turns on rollers instead of sliding on wooden blocks.
Meal floor: the floor beneath the stone floor, where the meal is stored after grinding.
Middlings: timbers which pass through the canister, having a sail bolted to each end. (also Stocks).
Nether stone: the lower, fixed, millstone. (also Bed stone).
Overdrift: stones driven from above.
Patent sails: automatically regulated shuttered sails.
Peak stones: millstones from the Peak District in Derbyshire.
Petticoat: wooden extension below a post-mill body or tower-mill cap to prevent rain entering the gap.
Piers: brick supports on which the ends of the cross-trees of a post-mill rest.
Pointing lines: ropes which enable a common sail to be adjusted from the ground or from a stage.
Poll-end: cast-iron socket at the end of the windshaft, which holds the stocks. (also Canister).
Post: heavy upright timber which supports the body of a post-mill.
Quant: spindle which carries the stone nut when the stones are overdrift.
Quarter bars: diagonal timbers which hold the post in a vertical position.
Rack: a gear round the top of a smock or tower-mill, enabling the cap to be turned by means of a gear system.
Roller-blind sails: similar to patent sails, but with roller-blinds replacing the shutters.
Round-house: circular brick building with a wooden roof, enclosing the substructure of a post-mill.
Runner stone: the upper, rotating, millstone.
Sack hoist: machinery geared to the main shaft to enable sacks of grain to be hauled up the mill.
Sail bars: the cross-bars of a sail frame.
Sail rods: connecting the shutter bars, enabling them to be controlled simultaneously.
Sheers: the two main horizontal timbers under the body of a post-mill.
Shoe: wooden trough below the hopper which directs the flow of grain into the eye of the runner stone.
Shutters: adjustable vanes in spring and patent sails.
Side girts: horizontal timbers fixed to the ends of the crown tree in a post-mill.
Spider: mechanism which enables all the sail rods to be controlled simultaneously by the striking rod.
Spill the wind: to let the wind pass through the sails by opening the shutters.
Spring sails: shuttered sails controlled by a pre-set spring device on each sail.
Stage: gallery round the outside of a mill.
Stocks: timbers which pass through the canister, having a sail bolted to each end. (also Middlings).
Stone floor: the floor of a mill on which the millstones are situated.
Stone nut: gear on the stone spindle, which engages with the great spur wheel.
Stone spindle: spindle which carries the runner stone.
Storm hatch: door at the top of a mill, giving access to the sails.
Striking rod: rod passing through the centre of the windshaft, enabling the shutters or roller-blinds to be adjusted while the sails are turning.
Substructure: the assembly of post, cross-trees, and quarter-bars, supporting a post-mill.
Sweeps: sails.
Tail: rear end of a post-mill.
Tail pole: beam extending to the ground behind a post-mill body, enabling it to be turned into the wind.
Tail winded: when the wind blows on the back of the sails.
Tentering: adjustment of the distance between the millstones.
Truck wheels: centring wheels fixed to the inside of a smock or tower-mill cap.
Underdrift: stones driven from below.
Vat: wooden casing enclosing the millstones.
Wallower: the first gear on the main upright shaft, engaging with the brake wheel.
Weather: the angle of twist on a sail.
Weather beam: main front beam in a post-mill, bearing the windshaft. (also Breast Beam).
Whips: the 'backbones' of the sails, which are bolted to the stocks.
Windshaft: shaft carrying the sails and the brake wheel.
Winding: turning the mill or cap to face into the wind.
Wire machine: device for separating the flour from the bran after grinding.

Bibliography

I. *Books and pamphlets about windmills in general, and also with relevance to sites in the Lambeth area.*

B1 Annual reports of the Society for the Protection of Ancient Buildings (Wind and Watermill Section), from 1930.

B2 British windmills and watermills, by Charles P. Skilton. Collins, 1947. 48p.

B3 Brixton windmill. Greater London Council, 1967. 8p.

B4 Discovering windmills, by John N. T. Vince. 2nd ed. Shire Publications, 1969. 64p.

B5 The Dutch windmill, by Frederick Stokhuyzen. C. A. J. van Dishoeck, n.d. (1962). 128p.

B6 England of the windmills, by Stuart P. B. Mais. J. M. Dent, 1931. 267p.

B7 The English windmill, by Rex Wailes. Routledge and Kegan Paul, 1954. 246p.

B8 English windmills, vol.1, by Marjorie I. Batten. Architectural Press, 1930. 128p.

B9 English windmills, vol.2, by Donald Smith. Architectural Press, 1932. 152p.

B10 History of corn milling, by Richard Bennett and John Elton. Simkin Marshall. 4 vols.
vol.1: Handstones, slave and cattle mills. 1898.
vol.2: Watermills and windmills. 1899.
vol.3: Feudal laws and customs. 1900.
vol.4: Some feudal mills. 1904.

B11 In search of English windmills, by Robert Thurston Hopkins and Stanley Freese. Cecil Palmer, 1931. 275p.

B12 Old English mills and inns, by Robert Thurston Hopkins. Cecil Palmer, 1927. 276p.

B13 Old watermills and windmills, by Robert Thurston Hopkins. Philip Allan, n.d. (1930). 245p.

B14 Transactions of the Newcomen Society, from 1920.

B15 Windmills, by Frank Brangwyn and Hayter Preston. John Lane, 1923. 126p.

B16 Windmills, by James Laver and Rex Wailes. Society for the Protection of Ancient Buildings, n.d. 22p.

B17 Windmills and millwrighting, by Stanley Freese. 2nd ed. David & Charles, 1971. 188p.

B18 Windmills in England: a study of their origin, development and future, by Rex Wailes. Architectural Press, 1948. 48p.

B19 The windmills of Surrey and inner London, by Kenneth G. Farries and M. T. Mason. Charles Skilton, 1966. 276p.

II. *Other books and pamphlets with relevance to windmills in the Lambeth area.*

B20 The Chronicles of Clapham, being a selection from the reminiscences of Thomas Parsons, by John H. Michael Burgess. Ramsden Press, 1929. 121p.

B21 Clapham: old yet ever new, by John R. Battley. Westminster City Publishing Co., 1937. 60p.

B22 Historic Battersea. Battersea Public Libraries, 1965. 20p.

B23 Historic Battersea, by Sherwood Ramsay. G. Rangecroft, 1913. 116p.

B24 History and antiquities of the county of Surrey, by Owen Manning and William Bray. John White, 1804. 3 vols.

B25 History and antiquities of the parish of Lambeth, by Thomas Allen. J. Allen, 1826. 458p.

B26 London memories: social, historical, and topographical, by Charles William Heckethorn. Chatto & Windus, 1900. 374p.

B27 London south of the Thames, by Sir Walter Besant. Adam & Charles Black, 1912. 372p.
B28 Magna Britannia and Hibernia: a new survey of Great Britain, by Thomas Cox and Anthony Hall. 1720-31. 6 vols.
B29 Natural history and antiquities of the county of Surrey, by John Aubrey. E. Curll, 1718. 5 vols.
B30 Old and new London: a narrative of its history, its people, and its places. vol.6: The southern suburbs, by Edward Walford. Cassell, Petter, Galpin & Co., n.d. (1898). 636p.
B31 Our lady of Batersey, by John G. Taylor. George White, 1925. 442p.
B32 Picturesque Norwood: reminiscences of Norwood, delivered by Arthur A. Saward at the Royal Normal College, Westow Street, Upper Norwood, under the auspices of the Upper Norwood Literary and Scientific Society, Wednesday, 10 April, 1907. J. A. Squire, 1907. 80p.
B33 Printed maps of London *c.*1553-1850, by Ida Darlington and James Howgego. George Philip, 1964. 257p.
B34 The Shires of England and Wales described by Christopher Saxton, revised by Philip Lea. n.d. (1690).
B35 Some glimpses of Norwood, by Alan R. Warwick. The Norwood Society, 1961. 23p.
B36 Survey of London, vol XXIII: South Bank and Vauxhall (The parish of St. Mary Lambeth, part 1). London County Council, 1951. 170p.
B37 Survey of London, vol. XXVI: The parish of St. Mary Lambeth (part 2: Southern area). London County Council, 1956. 226p.
B38 Victoria history of the county of Surrey. Constable, 1902. 4 vols.

References

N.B. Reference numbers in the text preceded by the letter B refer to publications listed in the Bibliography.

1. An Experimental enquiry concerning the natural powers of water and wind to turn mills and other machines depending on a circular motion (reprinted from the *Philosophical Transactions of the Royal Society*). London, 1760. British Museum.
2. Surrey Collection, Minet Public Library.
3. from the collection of Sir Ambrose Heal; photographed by Mr M. T. Mason.
4. National Grid reference TQ 30838035 (approx.).
5. British Museum Map Room: Crace Collection no. 35.
6. Photograph in the possession of Mr K. G. Farries.
7. Victoria and Albert Museum Print Room, ref. E1511-1931.
8. do., ref. E1510-1931.
9. (i) British Museum Print Room: Crace Collection no. 83.
   (ii) do., no. 35.
   (iii) do., no. 70.
10. Guildhall Library: information supplied by Mr K. G. Farries.
11. British Museum Print Room: Crace Collection no.73.
12. British Museum Map Room: Crace Collection no.143.
13. do., no.89.
14. do., no.58.
15. do., no.90.
16. National Grid reference TQ 31407995.
17. Duchy of Cornwall Office.
18. British Museum Map Room: Crace Collection no.173.
19. Surrey Collection, Minet Public Library.
20. British Museum Map Room: Crace Collection no.184.
21. Greater London Council Print Collection.
22. in *Illustrations of Surrey*, vol.2. Surrey Collection, Minet Public Library.
23. British Museum Print Room: Crace Collection no.97.
24. Surrey Collection, Minet Public Library.
25. British Museum Map Room.
26. (i) British Museum Map Room: Crace Collection no. 215.
   (ii) do., no.199.
27. In the possession of Mr Rex Wailes (information supplied by Mr K. G. Farries).
28. National Grid reference TQ 30877893.
29. British Museum Print Room: Crace Collection no.90.
30. do., no.87.
31. British Museum Map Room.
32. Surrey Collection, Minet Public Library.
33. (i) British Museum Map Room: Crace Collection no. 140.
   (ii) Surrey Collection, Minet Public Library.
34. Surrey Collection, Minet Public Library.
35. Greater London Council Record Office.
36. British Museum State Paper Room.
37. Guildhall Library.
38. Surrey Collection, Minet Public Library.
39. do.
40. Lambeth Palace Library.
41. British Museum Print Room: Crace Collection no.98.
42. Duchy of Cornwall Office.
43. Surrey Collection, Minet Public Library.
44. information supplied to Mr M. T. Mason by Mr H. E. S. Simmonds.

45. Surrey Collection, Minet Public Library.
46. photocopy in the Surrey Collection, Minet Public Library.
47. Surrey Collection, Minet Public Library.
48. Department of the Environment Library, Lambeth Bridge House: Mayson Beeton Collection no. XLI/4.
49. published in *Country Life*, 3 April 1958 (cutting in the possession of Mr K. G. Farries).
50. British Museum Map Room: Crace Collection no.253.
51. British Transport Historical Records.
52. photocopy in the Surrey Collection, Minet Public Library.
53. National Grid reference TQ 30707876.
54. Surrey Collection, Minet Public Library.
55. British Museum Map Room: Crace Collection no.173.
56. Surrey Collection, Minet Public Library.
57. do.
58. British Museum Map Room: Crace Collection no.265.
59. Surrey Collection, Minet Public Library.
60. British Museum Map Room: Crace Collection no.249.
61. Surrey Collection, Minet Public Library.
62. Guildhall Library.
63. Duchy of Cornwall Office.
64. National Grid reference TQ 30987870.
65. Surrey Collection, Minet Public Library.
66. (i) reprinted in *Atlas of London and the London Region*. Pergamon Press, 1968. Lambeth Central Reference Library.
(ii) British Museum Map Room: Crace Collection no. 140.
67. British Museum Map Room.
68. (i) Department of the Environment Library, Lambeth Bridge House: Mayson Beeton Collection.
(ii) British Museum Map Room: Crace Collection no. 168.
69. British Museum Map Room: Crace Collection no.159.
70. do., no.136.
71. Greater London Council Map Collection.
72. Guildhall Library.
73. Greater London Council Print Collection.
74. Guildhall Library.
75. Surrey Collection, Minet Public Library.
76. e.g. *Brixton Free Press*, 10 May 1912. Lambeth Central Reference Library.
77. Guildhall Library.
78. Battersea District Library.
79. British Museum Map Room: Crace Collection no.71.
80. Surrey Collection, Minet Public Library.
81. National Grid reference TQ 30007787.
82. Greater London Council Record Office.
83. exhibited 12 November to 31 December 1969 by Messrs Leger Galleries Ltd., Old Bond Street.
84. in the possession of Lord Spencer.
85. Battersea District Library.
86. Department of the Environment Library, Lambeth Bridge House: Mayson Beeton Collection no. XL/37.
87. reproduced on a Christmas card in the possession of Mr K. G. Farries: the original painting is stated to be in the Ante-room, Royal Army Medical Corps HQ Mess.
88. Victoria and Albert Museum Print Room (box no. U.12a).
89. Surrey Collection, Minet Public Library.
90. do.
91. British Museum Map Room: King's Topographical Collection no. XLI-27.K.
92. Guildhall Library.
93. do.
94. do.
95. Surrey Collection, Minet Public Library.
96. British Museum Print Room, ref. 1950-4-1-8; Imperial.
97. Brighton Art Gallery.
98. British Museum Map Room: Crace Collection no.249.
99. Surrey Collection, Minet Public Library.
100. Surrey Record Office, Kingston-upon-Thames.
101. Surrey Collection, Minet Public Library.
102. Library of the Royal Society.
103. Surrey Collection, Minet Public Library.
104. do.
105. Battersea District Library.
106. National Grid reference TQ 29027480.
107. Surrey Collection, Minet Public Library.
108. National Grid reference TQ 30467436.
109. *South London Press*, 26 June 1964. Lambeth Central Reference Library.
110. Surrey Collection, Minet Public Library.
111. do.
112. photocopy in the Surrey Collection, Minet Public Library.
113. Kent County Council Archives Office, Maidstone.
114. *The Times*, 5 December 1836. (microfilm in Lambeth Central Reference Library.)
115. copy in the Surrey Collection, Minet Public Library.
116. photocopy in the Surrey Collection, Minet Public Library.
117. Surrey Collection, Minet Public Library.
118. Lambeth Central Reference Library.
119. *The Lambethan*, no.3 (Spring 1964). Lambeth Central Reference Library.
120. documents in the Surrey Collection, Minet Public Library.
121. 7 August 1913.
122. 13 November 1913.
123. in the possession of the Greater London Council Historic Buildings Section.
124. deed in the Surrey Collection, Minet Public Library.
125. in the files of the Society for the Protection of Ancient Buildings.
126. obituary notice in the *Brixton Free Press*, 10 May 1935. Lambeth Central Reference Library.
127. press cuttings in the possession of the Greater London Council Surveyor, Lambeth District.
128. information from the Society for the Protection of Ancient Buildings and the Historic Buildings Section of the Greater London Council.
129. minutes of meetings in the possession of the Society for the Protection of Ancient Buildings.
130. Lambeth Borough Council minutes, 22 September 1954. Lambeth Central Reference Library.
131. information from the Greater London Council Historic Buildings Section.
132. *Streatham News*, 6 August 1954. Surrey Collection, Minet Public Library.
133. do., 13 August 1954.
134. do., 13 November 1959.
135. information from the Greater London Council Historic Buildings Section.
136. do.
137. report submitted to the London County Council by Mr Rex Wailes, 11 December 1958.
138. National Grid reference TQ 30327369.
139. view entitled 'Near Brixton'. Guildhall Library.
140. Surrey Collection, Minet Public Library.
141. G. Harley: 'Near Brixton Hill—windy day', in Lysons' *Environs of London*, vol.6: supplement (Grangerised). Guildhall Library.
142. Surrey Collection, Minet Public Library.
143. photocopy in the Surrey Collection, Minet Public Library.
144. *Illustrated News of the World*, 4 September 1858. British Museum Newspaper Library, Colindale.
145. Surrey Collection, Minet Public Library.
146. Public Record Office.
147. Surrey Collection, Minet Public Library.
148. do.
149. British Museum Map Room.
150. (i) Surrey Collection, Minet Public Library.
(ii) do.
(iii) British Museum Map Room: Crace Collection no. 241.
151. Surrey Collection, Minet Public Library.

152. National Grid reference TQ 31977323.
153. British Transport Historical Records.
154. Greater London Council Map Collection.
155. Surrey Record Office, Kingston-upon-Thames.
156. National Grid reference TQ 31857100.
157. Surrey Collection, Minet Public Library.
158. do.
159. all do.
160. do.
161. British Museum Map Room: Crace Collection no.9.
162. do., no.10.
163. Surrey Collection, Minet Public Library.
164. Croydon Central Reference Library.
165. Public Record Office.
166. Croydon Central Reference Library.
167. Public Record Office.
168. do.
169. National Grid reference TQ 33487068.
170. Public Record Office.
171. Kent County Council Archives Office, Maidstone.
172. information supplied by Mr Alan R. Warwick.
173. Guildhall Library: information supplied by Mr K. G. Farries.

Index